SpringerBriefs in Electrical and Computer Engineering

Signal Processing

Series editors

Woon-Seng Gan, Singapore, Singapore
C.-C. Jay Kuo, Los Angeles, USA
Thomas Fang Zheng, Beijing, China
Mauro Barni, Siena, Italy

Thomas Fang Zheng · Lantian Li

Robustness-Related Issues in Speaker Recognition

 Springer

Thomas Fang Zheng
Tsinghua National Laboratory
  for Information Science and Technology,
  Division of Technical Innovation
  and Development, Department
  of Computer Science and Technology
Center for Speech and Language
  Technologies, Research Institute
  of Information Technology,
  Tsinghua University
Beijing
China

Lantian Li
Tsinghua National Laboratory
  for Information Science and Technology,
  Division of Technical Innovation
  and Development, Department
  of Computer Science and Technology
Center for Speech and Language
  Technologies, Research Institute
  of Information Technology,
  Tsinghua University
Beijing
China

ISSN 2191-8112   ISSN 2191-8120 (electronic)
SpringerBriefs in Electrical and Computer Engineering
ISSN 2196-4076   ISSN 2196-4084 (electronic)
SpringerBriefs in Signal Processing
ISBN 978-981-10-3237-0   ISBN 978-981-10-3238-7 (eBook)
DOI 10.1007/978-981-10-3238-7

Library of Congress Control Number: 2017937265

Printed on acid-free paper

This Springer imprint is published by Springer Nature
The registered company is Springer Nature Singapore Pte Ltd.
The registered company address is: 152 Beach Road, #21-01/04 Gateway East, Singapore 189721, Singapore

Preface

Speaker recognition (known as voiceprint recognition in industry) is the process of automatically identifying or verifying the identity of a person from his/her voice, using the characteristic vocal information included in speech. It has gained great popularity in a wide range of applications in recent years. It enables the systems to use a person's voice to control the access to restricted services (such as automatic banking services), information (depending on the user's access right), or area (government or research facilities). It also allows the detection of speakers, such as voice-based information retrieval within audio archives, recognition of perpetrator in forensic analysis, and personalization of user devices.

After decades of research, current speaker recognition systems have achieved rather satisfactory performance. However, critical robustness issues still need to be addressed especially in practical situations. In this book, we provide an overview of technologies dealing with robustness-related issues in automatic speaker recognition. We categorize the robustness issues into three categories: environment-related, speaker-related, and application-oriented. For each category, we present current hot topics, the state-of-the-art technologies, and potential future research focuses.

Beijing, China

Thomas Fang Zheng
Lantian Li

Acknowledgements

This work was supported by the National Natural Science Foundation of China under Grant No. 61271389/61371136/61633013, and the National Basic Research Program (973 Program) of China under Grant No. 2013CB329302.

Contents

Chapter 1
Speaker Recognition: Introduction

In the ancient war times, officers and soldiers could recognize one friend or foe through the predetermined password(s). In real life, we human are able to get in and out of a house using keys or e-cards. While surfing the Internet, the user logins in websites or mail servers with his/her account and password. Even the electronic payment can be verified by a dynamic verification code sent to the user. All of the foresaid password, key, e-card, and dynamic verification code represent the identity information of the user. However, with the rapid development of technology and Internet, these traditional identity authentication ways cannot meet the needs for protecting users' personal information and properties. The password is liable to be leaked, key and e-cards lost and copied, account and password forgotten and attacked, and dynamic verification codes intercepted. These potential safety hazards lead to various accidents and troubles.

Nowadays accounts and passwords are ubiquitous, and humans are always bothered by forgetting or losing their passwords. Undoubtedly, biometric recognition is much more convenient and efficient as an alternative authentication method. Biometric recognition is a kind of automated technologies for measuring and analyzing an individual's physiological or behavioral characteristics, and can be used to verify or identify an individual. Biometrics [1] is instinctive and refers to metrics related to human characteristics, which can be applied in identity authentication anytime and anywhere. Obviously, with biometrics, one will not worry about forgetting and losing the passwords. Additionally, it has many advantages, including high security, unnecessity to remember, difficulty to be transformed or stolen, convenience, and efficiency. It was reported that the text-oriented password will become history and various traditional passwords will disappear, and all of them will be replaced by biometric recognition technology.

Biometrics can be divided into two categories: physiological characteristics and behavioral characteristics. Generally, the former consists of fingerprint, palm print, hand shape, face, iris, retina, vein, flavor, vascular pattern, DNA and so on, while the latter includes voiceprint, signature, gait, heart-beat and so on. Biometrics represents the inherent characteristics of a person, and has the property of universality,

© The Author(s) 2017

T.F. Zheng and L. Li, *Robustness-Related Issues in Speaker Recognition*,
SpringerBriefs in Signal Processing, DOI 10.1007/978-981-10-3238-7_1

uniqueness, stability and non-reproducibility. Yet in some practical applications, biometrics has its own limitations. In some cases, fingers and palms are exuvial, the authentication will be hard to realize. Outlaws can conceal its real identities to escape from justice by wearing fingerprint caps. Iris recognition technology also requires expensive camera focus and appropriate light source. Retina recognition technology needs laser irradiation on the back of the eyeballs to obtain the characteristics of retina, which will affect the health of user's eyes. Besides, the usability of retinal recognition technology is not quite smooth, and the cost of practical application is very high. In addition, there are also lots of risks in face recognition technology. Some spoofing methods, such as real-time facial reenactments, make it easier to attack face recognition system, weakening the non-reproducibility.

Compared with other biometrics, the voiceprint, belonging to the behavioral characteristics category, has the following advantages [2]:

1. Speech signal, which contains a series of voice characteristics, is easy and natural to obtain. And it is much easier for users to accept because the process of speech collection involves a little private conceal information.
2. The speech collecting devices are cheap and easy to use, for example, a microphone is enough. In many real scenarios, some communication equipments (such as telephone and mobile phone) can be used for speech collection, and there is no need for extra devices.
3. In combination with speech recognition technology, the dynamic voiceprint password can be applied, and users will not worry about the password being forgotten, lost, stolen, and fake recordings, thus it is suitable for remote authentication.

With these advantages, speaker recognition or voiceprint recognition, has gained a wide range of applications, such as access control, transaction authentication, voice-based information retrieval, recognition of perpetrator in forensic analysis, and personalization of user devices etc. This chapter describes speaker recognition in details, including its basic concepts, development history, typical system framework, categories and performance evaluations.

1.1 Basic Concepts

Spoken language is the most natural way we humans communicate with each other. There is rich information conveyed in spoken language, including language information (linguistic contents, accent etc.), speaker information (identity, emotion, physiological characteristics etc.), environmental information (background, channel etc.), and so on [3]. We humans can effortlessly decode most of such information although such rich information is encoded in a complex form. This ability of human has inspired a lot of research to automatically extract and process the richness of information in spoken language. As shown Fig. 1.1, many research tasks have been studied from different perspectives of spoken language, such as

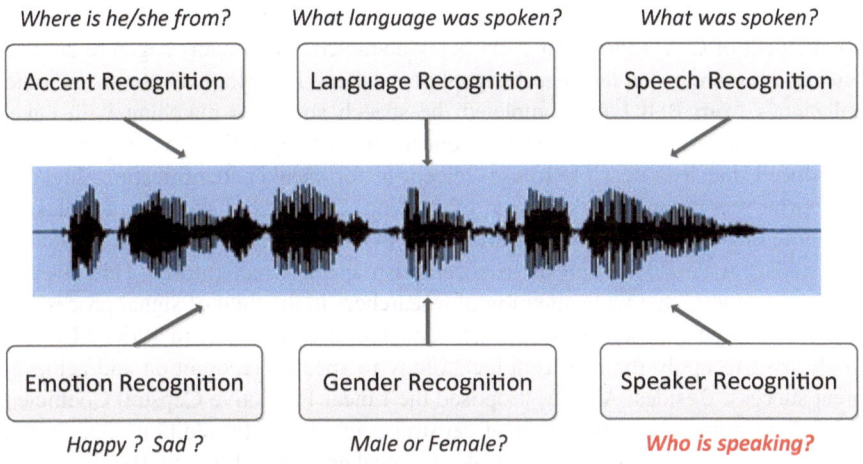

Fig. 1.1 Information contained in spoken language

accent recognition, language recognition, speech recognition, emotion recognition, gender recognition and speaker recognition.

It can be seen that the spoken language from a human voice involves all kinds of information. Among all the information, the voiceprint is a set of measurable characteristics of a human voice that uniquely symbolizes and identifies an individual. It belongs to the kind of behavioral characteristics. And the voiceprint (speaker) recognition refers to recognizing a person's identity from his/her voiceprint.

Research studies have suggested that the voiceprint of a person has the property of uniqueness and stability, which can stay relatively stable and is not easy to change especially in adulthood. Besides, researchers also believe that no two individuals sound identical because the shapes and sizes of their spoken organs, such as vocal tract, larynx, lungs, nasal cavity and other parts of their voice production organs, are quite different. In addition to these physiological differences, each speaker has his/her own characteristics or manner of speaking, including the use of a particular accent, rhythm, intonation style, pronunciation pattern, choice of vocabulary and so on. State-of-the-art speaker recognition systems use a number of these features in parallel, attempting to cover these different aspects and employing them in a complementary way to achieve more accurate recognition.

1.2 Development History

As the saying goes that a bird may be known by its song, people can distinguish who is speaking by his/her aural systems. Back to 1660s, the identity authentication by the human voice was firstly applied in a trial about the death of Charles I in Britain.

The research on speaker recognition was started from the 1930s. In 1937, with the incident of C.A. Lindbergh's son being abducted, we humans began to conduct scientific researches on the speech signal of speakers. In 1945, L.G. Kesta and other colleagues from Bell Labs completed the speech spectrum matching with naked eyes, and they firstly presented the concept of 'voiceprint'. Then in 1962, they introduced that it is possible to use voiceprint for speaker identification. In 1966, voiceprint was regarded as one kind of evidence in courts of the United States. At the same time, S. Pruzanshy from Bell Labs [4] put forward the template matching and statistical variance analysis to accomplish speaker recognition. This research achievement attracted wide attention of researchers in the field of signal processing, and the upsurge of research on speaker recognition was set off. In 1969, J.E. Luck firstly tried to apply the cepstrum technology to speaker recognition and achieved great success. Besides, Atal [5] proposed the Linear Predictive Cepstral Coefficient (LPCC) to improve the precision of cepstral coefficients. In addition, Doddington [6] came up with the resonance peaks in speaker verification. In 1972, Atal [7] presented the concept of fundamental frequency for speaker recognition.

From 1970s to the end of 1980s, research on speaker recognition was concentrated on acoustic feature extraction and some template-matching methods. Researchers had come up with several feature parameters, including the Perceptual Linear Predictive (PLP) [8], Mel-frequency Cepstral Coefficient (MFCC) [9]. Besides, models such as Dynamic Time Warping (DTW) [10], Vector Quantization (VQ) [11], Hidden Markov Model (HMM) [12], Artificial Neural Network (ANN) [13] have been widely used in the field of speaker recognition, and became the main technologies of speaker recognition.

Since 1990s, especially after the detailed introduction of Gaussian mixture model (GMM) by Reynolds [14], GMM had rapidly become a mainstream model for text-independent speaker recognition, due to its advantage of easy catch, flexibility, high efficiency and good robustness. In 2000, Reynolds [15] brought up the GMM-UBM (Gaussian mixture model—Universal background model), which had made great contribution to making speaker recognition technology from lab experiment to practical use. The framework of GMM-UBM system is as shown Fig. 1.2. Based on this model, some derived models appeared, such as GMM-HMM and GMM-SVM.

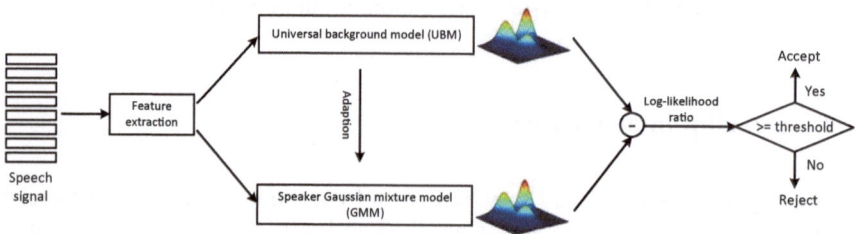

Fig. 1.2 The framework of a GMM-UBM system

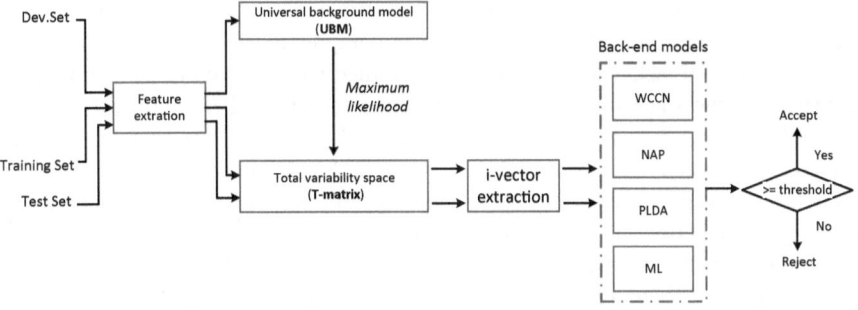

Fig. 1.3 The framework of an i-vector system

In the beginning of the 21st century, according to the traditional GMM-UBM, P. Kenny, N. Dehak and other researchers proposed the Joint Factor Analysis (JFA) [16] and the i-vector [17] models. In these models, overcoming the limitations of mutual independence of Gaussian components in GMM-UBM system, the speaker models were embedded into a low-dimensional subspace, and improved the system performance. Accompanied with these models, various back-end techniques had also been proposed to promote the discriminative capability for speakers, such as within-class covariance normalization (WCCN) [18], nuisance attribute projection (NAP) [19], probabilistic linear discriminant analysis (PLDA) [20–22] and metric learning (ML) [23], etc. These methods had been demonstrated to be highly successful.

As an example, the framework of an i-vector system is shown in Fig. 1.3.

Before 2010, most of the successful approaches to speaker recognition were based on generative models with unsupervised learning, e.g., the GMM-UBM framework and the i-vector model. Despite the impressive success of these models, they still shared the intrinsic disadvantage of all unsupervised learning models: the goal of the model training was to describe the distributions of the acoustic features, instead of discriminating speakers. Inspired by the success of deep neural networks (DNNs) in speech recognition [24–26], DNN and its recurrent variant (recurrent neural networks, RNN) had been applied in speaker recognition and also achieved promising results. Based on these models, many intriguing structures were designed for deep feature learning or deep speaker embedding. For example, in [27, 28], DNNs trained for ASR (automatic speech recognition) were used to replace the UBM model to derive the acoustic statistics for i-vector models, and this phonetic-aware method could provide more accurate frame posteriors for statistics computation. In [29], a DNN was used to replace the PLDA to improve the discriminative capability of i-vectors. Besides, Ehsan [30] proposed a DNN structure for deep speaker feature learning. Followed by this work, a phonetic-dependent DNN structure [31] was studied by Li. In this model, phone posteriors were involved in the DNN input so that speaker discriminative features could be learned more easily by alleviating the impact of phone variation. More interestingly, research studied that there was a close correlation between automatic speech recognition and speaker recognition. And the 'multi-task learning' made the joint training and inference for correlated task possible. For example, Chen and colleagues [32] found that phone recognition and grapheme recognition can be treated as two correlated tasks, and a DNN model trained with the tasks as objectives outperforms the ones trained only with phone targets. Other multi-task learning research work was reviewed in [33]. Moreover, recently, Tang and Li [34] presented a collaborative learning structure based on long short-term memory (LSTM) models. The main idea was to merge task-specific neural models with inter-task recurrent connections into a unified model. This model fits well with the joint training of speech and speaker recognition. In this scenario, the speech content and speaker identity were produced at each frame step by the speech and speaker components, respectively. By exchanging these bits of information, performances of both speech and speaker recognition were sought to be improved. This led to a joint and simultaneous learning for the two tasks. Figure 1.4 gives an example of this proposed multi-task recurrent model. In this model, the recurrent information is extracted from both the recurrent projection r_t and the non-recurrent projection p_t, and the information is used as additional input to the non-linear function $g(\cdot)$. Experimental results showed that the presented approach could learn speech and speaker models in a joint way and could improve the performance for both tasks. Furthermore, David [35] investigated an end-to-end text-independent speaker verification system. As shown in Fig. 1.5, the architecture consists of a deep neural network that takes a variable-length speech segment and maps it to a speaker embedding. The objective function separates same-speaker and different-speaker

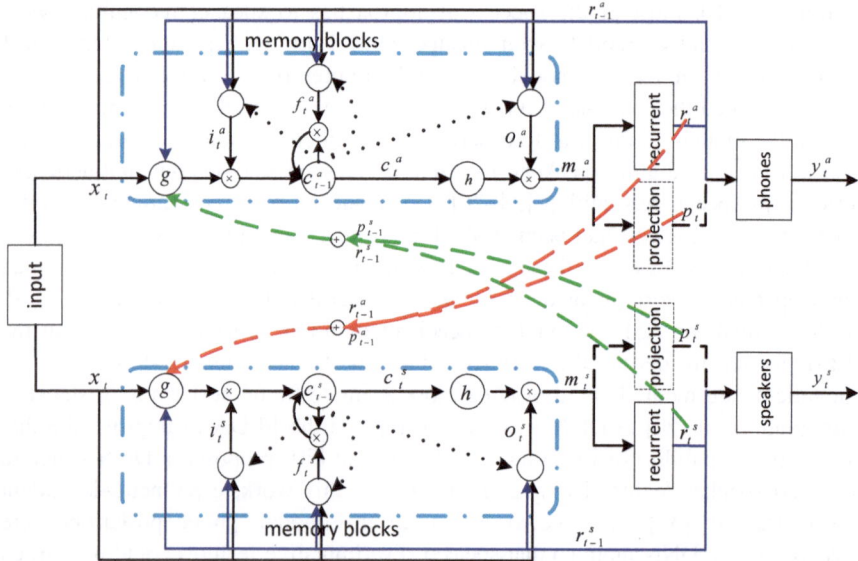

Fig. 1.4 The framework of multi-task recurrent models for speech and speaker recognition

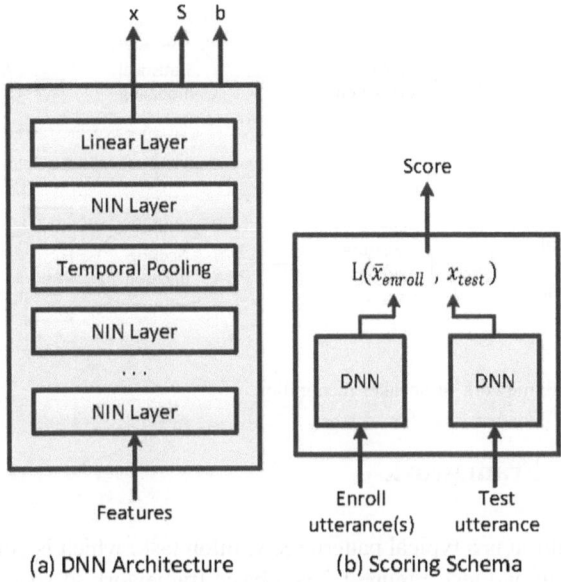

Fig. 1.5 The framework of end-to-end speaker verification system

pairs, and is reused during verification. In this model, both deep speaker embedding and discriminative training are combined together. The results showed that this end-to-end model delivered much better performance than the i-vector model especially on conditions of short utterances.

This section reviews the history of speaker recognition technologies as briefly summarized in Fig. 1.6. With the development of time, the speaker recognition systems have been improved continuously, and the application conditions have converted from the small-scale clean laboratory data into large-scale practical application data.

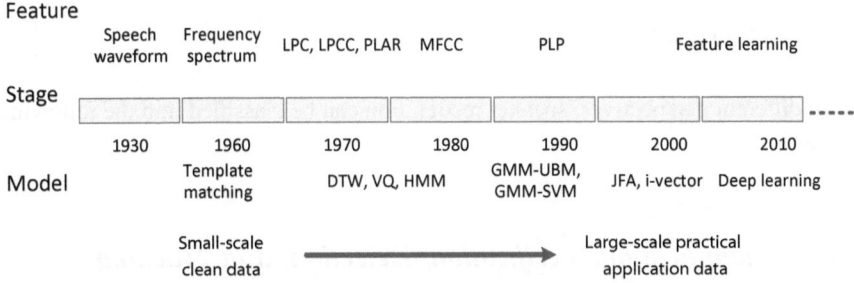

Fig. 1.6 Development history of speaker recognition technology

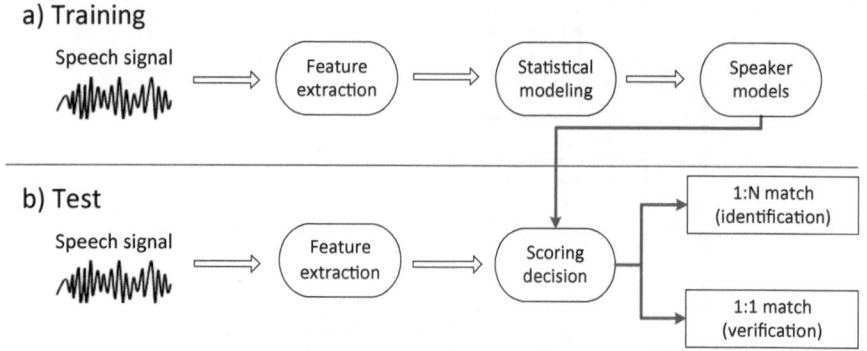

Fig. 1.7 System framework of speaker recognition

1.3   System Framework

Speaker recognition is a typical pattern recognition task, which is composed of two modules, training and test. Figure 1.7 is a basic framework of speaker recognition system.

The training, as shown in Fig. 1.2a, is also called registration or enrollment, in which a user enrolls by providing voice samples to the system. It consists of two steps. The first step is to extract features from the speech signal. The second step is to obtain a statistical model from the extracted features, referred to as speaker models. The test, as shown in Fig. 1.2b, is also called classification or recognition, in which a test voice sample is used by the system to measure the similarity between the user's voice and each of the previously enrolled speaker models so as to make a decision. In a speaker identification task, the system measures the similarity between the test sample and every stored speaker model, while in a speaker verification task, the similarity is measured only with the model of the claimed identity. The decision also differs across systems.

1.4   Categories

From different perspectives, speaker recognition can be classified into the following categories.

1.4.1   Identification, Verification, Detection, and Tracking

Basically, speaker recognition can be categorized into two fundamental categories [36–38]: identification and verification.

Speaker identification determines which identity in a specified speaker set is speaking given a speech segment. There are two modes of operation for speaker identification: open-set and close-set. In the close-set mode, the system assumes that the unknown speech samples must come from one in the set of enrolled speakers. During testing, a matching score is estimated for each enrolled speaker and the speaker corresponding to the model with the highest matching score is selected. In the open-set mode, the speaker can be inside or outside of the set of enrolled speakers, and those who are not enrolled should be rejected. This requires another model referred to as imposter model or background model, which is trained with data provided by other speakers different from the set of enrolled speakers. During testing, the matching score corresponding to the best speaker model is compared with the matching score estimated using the imposter model. Speaker identification performance tends to decrease as the population size increases. An important application of speaker identification technology is forensics analysis, identifying the suspects among a set of known criminals.

Speaker verification is also named as voice verification or authentication, which is to determine whether a claimed identity is speaking during a speech segment. It is a binary decision task. During testing, a verification score is estimated based on the claimed speaker's model. This verification score is then compared with a pre-defined threshold. If the score is higher than the threshold, the test is accepted, otherwise, rejected. Hence, the performance is quite independent of the population size, but it depends on the number of test utterances used to evaluate the system performance. This task can be used for security applications, such as controlling telephone access to banking services.

Additionally, there are another two kinds of speaker recognition, speaker detection (to determine whether a specified target speaker is speaking during a speech segment) and speaker tracking (to performing speaker detection as a function of time, giving the timing index of the specified speaker, known as *diarization*). The speaker diarization task can be called as a "*who spoken when*" recognition task. Different from either identification or verification, this task can apply to multi-speaker conversation scenarios. The system identifies the speaker changes and clusters the segments that belong to the same speaker. This task can be used for spoken document indexing and retrieval, meta data generation, etc.

1.4.2   Text-Dependent, Text-Independent and Text-Prompted

Automatic speaker recognition systems can also be classified according to the speech modality [37]: text-dependent, text-independent, or text-prompted.

In the text-dependent (TD) mode, the user is expected to say a pre-determined text for both training and test. Due to the prior knowledge (lexical content) of the spoken phrase, TD systems are generally more robust and can achieve good performance. However, there are cases when such constraints can be cumbersome or impossible to enforce. Besides, for TD systems, the ability against spoofing attack is very weak. Once the prospective imposters steal the text information, the systems will be easily broken up.

In the text-independent (TI) mode, there are no constraints on the text. Thus, the enrollment and test utterances may have completely different texts. For such cases, it is more convenient for users to operate. Unfortunately, since the content information is not used, there exists a distribution mismatch between enrollment and test due to the text variations, which leads to performance degradation.

Combining the advantages of the TD and TI modes, the text-prompted systems come into being. In the text-prompted (TP) mode, the text to speak is not fixed each time when use, but (randomly) prompted by the system. In such a system, the speech recognition as well as the speaker recognition will be performed normally, to recognize both the identity and the content information, which can be regarded as a kind of *"who spoke what"* recognition task. The TP system accepts an input utterance only when it judges that it was spoken by the claimed speaker and the content of it was the same as the prompted text. Both of them are indispensable. On the other hand, these prompted texts can be dynamic, and this property makes it have a good capability of anti-spoofing. Due to these advantages, the text-prompted speaker recognition systems have been the preferred alternative in many practical applications.

1.5 Performance Evaluations

In this section, we will introduce several evaluation metrics for speaker recognition, and they are slightly different for different categories. For a speaker verification task or an open-set speaker identification task, the Detection Error Trade-offs (DET) curve, Equal Error Rate (EER), and Detection Cost Function (DCF) are often used as the evaluation metrics, while for a close-set speaker identification task, different modes of the test set still have different evaluation metrics.

1.5.1 Evaluation Metrics for Verification or Open-Set Identification

1. Detection Error Trade-offs (DET) curve [39]

There are two important metrics for performance evaluation of speaker verification or open-set speaker identification systems, false acceptance rate (FAR), and false rejection rate (FRR). FAR is the measure of the likelihood that the system incorrectly accepts an identity claim from an impostor. FRR is the measure of the likelihood that the system incorrectly rejects an identity claim from a valid speaker. The FAR and FRR can be defined as:

$$\text{FAR} = \frac{\text{the number of false accepted imposter trials}}{\text{the total number of imposter trials}} \times 100\%$$

$$\text{FRR} = \frac{\text{the number of false rejected target trials}}{\text{the total number of target trials}} \times 100\%$$

Both the FAR and FRR depend on the threshold θ used in the decision-making process, where θ is a trade-off between the two types of errors. With a low threshold, the system tends to easily accept every identity claim thus making few false rejections but lots of false acceptances. On the contrary, if the threshold is set to some high value, the system will easily reject every claim and make very few false acceptances but a lot of false rejections.

Obviously, the two error rates are functions of the decision-making threshold θ. Therefore, it is possible to represent the performance of a system by plotting FAR as a function of FRR. This curve is known as the detection error trade-offs (DET) curve. For a speaker verification system, the horizontal axis represents the FAR, and the vertical axis demonstrates the FRR. Adjusting the decision threshold θ, the variation trends between FAR and FRR can be plotted in a DET curve. For better distinguishing different well performing systems, a normal deviate scale will be used for both axes, resulting in a pseudo-linear curve with a slope equal to -1. The better the system is, the closer to the origin zero the curve will be. Figure 1.8 shows an example of DET curve.

In the DET curve, the error rate value at the point where FAR equals FRR is called the Equal Error Rate (EER). Obviously, the smaller the EER value, the better

Fig. 1.8 Example of a DET curve

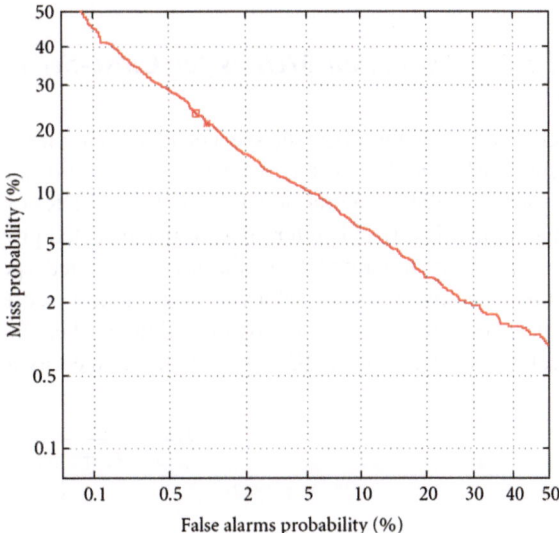

the system performance. It represents a balanced performance and is widely used as the evaluation metrics for speaker verification.

2. Detection Cost Function (DCF) [40]

In calculating the DET curve and EER, the evaluator effectively chooses optimal decision thresholds. However, in real applications, the proportion of targets and non-targets may be different from the proportion of the evaluation database. Moreover, the two types of errors may not have equally grave consequences. For example, for a fraud detection scenario, the cost of a false rejection (missing target) can be higher than the cost of a false acceptance, while for access control the cost of a false acceptance may overweight that of a false rejection in order for security. Therefore, it makes sense to take into account the two error rates weighted by their respective costs. Apply these weightings, one then arrives at a scalar performance measure, namely the expected cost of detection errors,

$$C_{\det}(FRR, FAR) = C_{FR} * FRR * P_{tar} + C_{FA} * FAR * (1 - P_{tar}).$$

This function has become known as the *detection cost function*. Here FRR and FAR are the two types of error rates as defined above, C_{FR} and C_{FA} are application dependent cost parameters, representing the cost weighting for FRR and FAR, respectively, and P_{tar} is the prior probability that a target speaker event occurs in the application. $C_{\det}$ has been used since the first NIST speaker recognition evaluation in 1996 as the primary evaluation measure, and with it, the cost parameters have been assigned values $C_{FR} = 10$, $C_{FA} = 1$ and $P_{tar} = 1\%$. Besides, the measure minimum $C_{\det}$, abbreviated as minDCF, is defined as the optimal value of $C_{\det}$ obtained by adjusting the decision threshold. Similar to EER, it is a summarization of DET curve and is a very important performance metric.

1.5.2   Evaluation Metrics for Close-Set Identification

Generally, for the open-set speaker identification system, it can still use the DET curve, EER or DCF as the evaluation metrics. For a close-set speaker identification system, it often uses the identification rate (IDR) or Top-N accuracy as the evaluation metrics. IDR is referred as to the expected proportion that the test utterances are correctly identified from the set of enrolled speakers. Usually, for each test utterance, the speaker achieving the highest score among the set of enrolled speakers is regarded as the identification speaker. In such a case, IDR can also be called Top-1 accuracy. IDR is formally defined as follows:

$$IDR = \frac{T_c}{T_c + T_i}$$

where the T_c and T_i are the numbers of correctly and incorrectly identified utterances, respectively. The evaluation metric of Top-N accuracy is defined as the proportion of the top-n hits in all test trials. Given a test utterance, if the target speaker is in the n-nearest candidates of the enrolled speaker set, then a top-n hit is obtained and this test trial is regarded as correctness.

References

1. Wikipedia. https://en.wikipedia.org/wiki/Biometrics
2. Zhang C (2014) Research on short utterance speaker recognition. Tsinghua University, Ph.D. Dissertation
3. Zheng TF, Jin Q, Li L et al (2014) An overview of robustness related issues in speaker recognition. Asia-Pacific Signal and Information Processing Association, 2014 Annual Summit and Conference (APSIPA). IEEE, pp 1–10
4. Furui S (2005) 50 years of progress in speech and speaker recognition. SPECOM 2005, Patras, pp 1–9
5. Atal BS, Hanauer SL (1971) Speech analysis and synthesis by linear prediction of the speech wave. J Acoust Soc Am 50(2B):637–655
6. Doddington GR, Flanagan JL, Lummis R C (1972) Automatic speaker verification by non-linear time alignment of acoustic parameters. U.S. Patent 3,700,815 [P], pp 10–24
7. Atal BS (1972) Automatic speaker recognition based on pitch contours. J Acoust Soc Am 52 (6B):1687–1697
8. Hermansky H (1990) Perceptual linear predictive (PLP) analysis of speech. J Acoust Soc Am 87(4):1738–1752
9. Vergin R (1999) O'shaughnessy D, Farhat A. Generalized mel frequency cepstral coefficients for large-vocabulary speaker-independent continuous-speech recognition. IEEE Trans Speech Audio Process 7(5):525–532
10. Sakoe H, Chiba S (1978) Dynamic programming algorithm optimization for spoken word recognition. IEEE Trans Acoust Speech Signal Process 26(1):43–49
11. Burton D, Shore J, Buck J (1983) A generalization of isolated word recognition using vector quantization. Acoustics, speech, and signal processing. IEEE international conference on ICASSP'83. IEEE vol 8, pp 1021–1024
12. Rabiner L, Juang B (1986) An introduction to hidden Markov models. IEEE ASSP Magazine 3(1):4–16
13. Jain AK, Mao J, Mohiuddin KM (1996) Artificial neural networks: a tutorial. Computer 29(3):31–44
14. Reynolds D (2015) Gaussian mixture models. Encyclopedia of biometrics, pp 827–832
15. Reynolds DA, Quatieri TF, Dunn RB (2000) Speaker verification using adapted Gaussian mixture models. Digit Signal Proc 10(1–3):19–41
16. Dehak N, Dumouchel P, Kenny P (2007) Modeling prosodic features with joint factor analysis for speaker verification. IEEE Trans Audio Speech Lang Process 15(7):2095–2103
17. Dehak N, Kenny P, Dehak R et al (2011) Front-end factor analysis for speaker verification. IEEE Trans Audio Speech Lang Process 19(4):788–798
18. Hatch AO, Kajarekar SS, Stolcke A (2006) Within-class covariance normalization for SVM-based speaker recognition. INTERSPEECH
19. Solomonoff A, Quillen C, Campbell WM (2004) Channel compensation for SVM speaker recognition. Odyssey, vol 4, pp 219–226
20. McLaren M, Van Leeuwen D (2011) Source-normalised-and-weighted LDA for robust speaker recognition using i-vectors. Acoustics, speech and signal processing (ICASSP), 2011 IEEE international conference on. IEEE, pp 5456–5459

21. Ioffe S (2006) Probabilistic linear discriminant analysis. European conference on computer vision. Springer, Berlin, pp 531–542
22. Prince SJD, Elder JH (2007) Probabilistic linear discriminant analysis for inferences about identity. Computer vision, 2007. ICCV 2007. IEEE 11th international conference on. IEEE, pp 1–8
23. Yang L (2007) An overview of distance metric learning. Proceedings of the computer vision and pattern recognition conference
24. Dahl GE, Yu D, Deng L et al (2012) Context-dependent pre-trained deep neural networks for large-vocabulary speech recognition. IEEE Trans Audio Speech Lang Process 20(1):30–42
25. Graves A, Jaitly N (2014) Towards end-To-end speech recognition with recurrent neural networks. ICML, vol 14, pp 1764–1772
26. Sak H, Senior AW, Beaufays F (2014) Long short-term memory recurrent neural network architectures for large scale acoustic modeling. INTERSPEECH, pp 338–342
27. Lei Y, Scheffer N, Ferrer L et al (2014) A novel scheme for speaker recognition using a phonetically-aware deep neural network. Acoustics, speech and signal processing (ICASSP), 2014 IEEE international conference on. IEEE, pp 1695–1699
28. Kenny P, Gupta V, Stafylakis T et al (2014) Deep neural networks for extracting baum-welch statistics for speaker recognition. Proc. Odyssey, pp 293–298
29. Wang J, Wang D, Zhu Z et al (2014) Discriminative scoring for speaker recognition based on i-vectors. Asia-pacific signal and information processing association, 2014 annual summit and conference (APSIPA). IEEE, pp 1–5
30. Variani E, Lei X, McDermott E et al (2014) Deep neural networks for small footprint text-dependent speaker verification. Acoustics, speech and signal processing (ICASSP), 2014 IEEE international conference on. IEEE, pp 4052–4056
31. Li L, Lin Y, Zhang Z et al (2015) Improved deep speaker feature learning for text-dependent speaker recognition. Signal and information processing association annual summit and conference (APSIPA), 2015 Asia-Pacific. IEEE, pp 426–429
32. Chen N, Qian Y, Yu K (2015) Multi-task learning for text-dependent speaker verification. Sixteenth annual conference of the international speech communication association, pp 185–189
33. Wang D, Zheng TF (2015) Transfer learning for speech and language processing. Signal and information processing association annual summit and conference (APSIPA), 2015 Asia-Pacific. IEEE, pp 1225–1237
34. Tang Z, Li L, Wang D et al (2016) Collaborative joint training with multi-task recurrent model for speech and speaker recognition. IEEE/ACM Transactions on Audio, Speech, and Language Processing
35. Snyder D, Ghahremani P, Povey D et al (2016) Deep neural network-based speaker embeddings for end-to-end speaker verification, 2016 IEEE Workshop on Spoken Language Technology
36. Furui S (1997) Recent advances in speaker recognition. Pattern Recogn Lett 18(9):859–872
37. Campbell JP (1997) Speaker recognition: a tutorial. Proc IEEE 85(9):1437–1462
38. Tranter SE, Reynolds DA (2006) An overview of automatic speaker diarization systems. IEEE Trans Audio Speech Lang Process 14(5):1557–1565
39. Martin A, Doddington G, Kamm T et al (1997) The DET curve in assessment of detection task performance. Proc of the European conference on speech communication and technology (Eurospeech 1997), Rhodes, Greece, vol 4, pp 1895–1898
40. The NIST year 2006 speaker recognition evaluation plan. http://www.itl.nist.gov/iad/mig/tests/sre/2006/sre-06_evalplan-v9.pdf

Chapter 2
Environment-Related Robustness Issues

In practical applications, many environment-related factors may influence the performance of speaker recognition. There is often *no prior* knowledge of these factors in advance, which makes the environment-related robustness issue more difficulty. In this chapter, three environment-related factors, background noise, cross channel and multiple-speaker, are summarized and their corresponding robustness issues are discussed.

2.1 Background Noise

The speech wave recorded in real environments often contains different types of background noises such as white noise, car noise, music etc. The background noise has adverse impact on speaker modeling and disturbs the evaluation testing, and so degrades the performance of speaker recognition system. The research on background noise robustness generally has four directions: speech enhancement, feature compensation, robust modeling, and score normalization.

2.1.1 Speech Enhancement

Despite the fact that the conventional and the state-of-the-art speech enhancement techniques have been gained satisfactory effects, employing signal-level enhancement has shown to be effective in improving speaker recognition in noisy environments. In [1], the subtractive noise suppression analysis was presented, and the spectral subtraction was proposed to suppress stationary noise from speech by subtracting the spectral noise bias calculated during non-speech activity and attenuate the residual noise left after subtraction. Since this algorithm resynthesizes a speech waveform, it can be used as pre-processing to a speaker recognition

© The Author(s) 2017
T.F. Zheng and L. Li, *Robustness-Related Issues in Speaker Recognition*,
SpringerBriefs in Signal Processing, DOI 10.1007/978-981-10-3238-7_2

system. In [2], different techniques were used to remove the effect of additive noise on the vocal source features WOCOR and the vocal track features MFCC. And a frequency-domain approach was proposed to denoise the residual signal and hence improve the noise-robustness of WOCOR. However, these methods do not perform well when noise is nonstationary. RASTA filtering [3] and cepstral mean normalization (CMN) [4] have been used in speaker recognition but they are mainly intended for convolutive noises. Inspired by auditory perception, computational auditory scene analysis (CASA) [5] typically segregates speech by producing a binary time-frequency mask. To deal with noisy speech, Zhao applied CASA separation and then reconstructed or marginalized corrupted components indicated by a CASA mask. It was further shown in [6] that these algorithms might either enhance or degrade the recognition performance depending on the noise type and the SNR level.

2.1.2 Feature Compensation

There are also algorithms to improve the system robustness in feature domain. In [7], 12 different short-term spectrum estimation methods were compared for speaker verification under the additive noise contamination. Experimental results conducted on the NIST 2002 SRE show that the spectrum estimation method has a large effect on recognition performance and the stabilized weighted LP (SWLP) and the minimum variance distortionless response (MVDR) methods can yield approximately 7 and 8% relative improvements over the standard DFT method in terms of EER. Lei et al. [8, 9] proposed a vector Taylor series (VTS) based i-vector model for noise-robust speaker recognition by extracting synthesized clean i-vectors to be used in the standard system back-end. This approach brought significant improvements in accuracy for noisy speech conditions. Martinez et al. [10] tried to model non-linear distortions in cepstral domain based on a nonlinear noise model in order to relate clean and noisy cepstral coefficients and help estimate a "cleaned-up" version of i-vectors. Moreover, to avoid the high computational load of the i-vector modelling in the proposed noisy environment, a simplified version is followed, where the sufficient statistics are normalized with their corresponding utterance-dependent noise adapted UBM.

2.1.3 Robust Modeling

The research on model robustness against noise usually adopts model compensation algorithms to decrease the mismatch between the test and the training utterances.

The parallel model combination (PMC) was first introduced in speech recognition [11] in advance of speaker recognition [12] by building a noisy model and using it to decode noisy test segments. This iterative method compensates additive and convolutive noises directly at the data level. The main advantages of this method are to allow the compensation of the noise presenting in both test and training data, to take into account the variance of the different noises, and to facilitate the use of delta coefficients.

2.1.4 Score Normalization

A robust back-end training called "multi-style" [13] was proposed as a possible solution to noise reduction in the score level. This method used a large set of clean and noisy data (affected with different noises and SNR levels) to build a generic scoring model. The obtained model gave good performance in general but was still suboptimal (for a particular noise) because of its generalization (the same system was used for all noises). Adding noisy training data in the current i-vector based approach followed by probabilistic linear discriminant analysis (PLDA) can bring significant gains in accuracy at various signal-to-noise ratio (SNR) levels. Besides, [14] proposed a method for determining the nuisance characteristics presenting in an audio signal. The method relied on the extraction of i-vectors over the signal, an approach borrowed from the speaker recognition literature. Given a set of audio classes in the training data, a Gaussian model was trained to represent the i-vectors for each of these classes. During recognition, these models were used to obtain the posterior probability of each class given the i-vector for a certain signal. This framework allowed for a unified way of detecting any kind of nuisance characteristic that was properly encoded in the i-vector used to represent the signal.

2.2 Channel Mismatch

Channel mismatch is another salient factor that influences the recognition performance. In real applications, speech utterances are often recorded with various types of microphones (such as desktop microphone and head phone), and these speech signals are changed in some degree due to different transmission channels. In every Speaker Recognition Evaluations organized by NIST [14], the channel mismatch issue was always regarded as one of the most important challenges. To encourage the research dealing with channel mismatch issues, different recording devices and transmission channels have been utilized in collecting the evaluation data [15, 16]. Nowadays, research dealing with the channel mismatch of speaker verification tasks can be categorized into three directions: feature transformation, model compensation, and score normalization.

2.2.1   Feature Transformation

CMS (Cepstral Mean Subtraction) [17] or Cepstral Mean Normalization (CMN) which subtracts the mean value of each feature vector over the entire utterance is the simplest and most commonly-used method for many speaker verification systems. The channel variations are considered to be stable over the entire utterance in these methods. Feature mapping [18] that maps features to a channel-independent feature space and feature warping which modifies the short-term feature distribution to follow a reference distribution are also effective methods but with more complex implementation.

2.2.2   Channel Compensation

SMS (Speaker Model Synthesis) is popular in GMM-UBM systems, which transforms models from one channel to another according to the UBM deviations between channels. Reference [19] proposed a novel statistical modeling and compensation method. In channel-mismatched conditions, the new approach uses speaker-independent channel transformations to synthesize a speaker model that corresponds to the channel of the test session. A cohort-based speaker model synthesis (SMS) algorithm, designed for synthesizing robust speaker models without requiring channel-specific enrollment data, was proposed in [20]. This algorithm utilized a priori knowledge of channels extracted from speaker-specific cohort sets to synthesize such speaker models. Besides, Ref. [21] explored techniques specific to the SVM framework in order to derive fully non-linear channel compensations.

Factor analysis is another model-level compensation method to analyze the discrimination of speaker models over different channels. [22] proposed a hybrid compensation scheme (both in the feature and the model domains). The implementation is simpler, as the target speaker model does not change over the verification experiment and the standard likelihood computation can be employed. In addition, while the classical compensation scheme brings a bias in scores (score normalization is needed to obtain good performance), this approach presents good results with native scores. Finally, the use of a SVM classifier with a proper supervector-based kernel is straightforward. JFA (Joint Factor Analysis) [23], a more comprehensive statistical approach, has gained much success in speaker verification. The speaker variations and channel (session) variations were modeled as independent variables spanning in a low-rank subspace, which defined the speaker-and channel-variations as two independent random variables following a priori standard Gaussian distributions. Then the factors were inferred the posterior probability of the speaker-and channel-variations from the given speech.

The i-vector method [24] assumes that the speaker and channel variations cannot be separated by JFA because the channel variation also contains speaker information. So in the i-vector method, a low-rank total variability space is defined to represent speaker-and channel-variations at the same time, and the speaker utterance is represented by an i-vector which is derived by inferring the posterior distribution of the total variance factor. There is no distinction between speaker effects and channel effects in GMM supervector space. Both speaker-and channel-variations are retained in i-vector. However, the total representation will lead to less discrimination among speakers due to channel variations. Therefore, many inter-channel compensation methods especially some popular discriminative approaches are employed to extract more accentuated speaker information. WCCN (With-in Class Covariance Normalization) [25] and LDA (Linear Discriminant Analysis) [24, 26] are both linear transformation to optimize the linear kernels. NAP (Nuisance Attribute Projection) [21] is to find the projection optimized by minimizing the difference among channels.

The most recent research focuses on the PLDA (Probabilistic Linear Discriminant Analysis) [27, 28], which can improve the performance of an i-vector system greatly. PLDA is a probabilistic version of LDA, and also is a generative model that utilizes a prior distribution on the speaker-and channel-variations. PLDA plus length normalization was reported to be most effective. The success of this model is believed to be largely attributed to two factors: one is its training objective function that reduces the intra-speaker variation while enlarges inter-speaker variation, and the other is the Gaussian prior that is assumed over speaker vectors, which improves robustness when inferring i-vectors for speakers with limited training data.

These two factors, however, are also two main shortcomings of the PLDA model. As for the objective function, although it encourages discrimination among speakers, the task in speaker recognition is to discriminate the true speaker and the imposters which is a binary decision rather than the multi-class discrimination in PLDA training. As for the Gaussian assumption, although it greatly simplifies the model, the assumption is rather strong and is not practical in some scenarios, leading to less representative models.

Some researchers have noticed these problems. For example, to go beyond the Gaussian assumption, Kenny [29] proposed a heavy-tailed PLDA which assumed a non-Gaussian prior over the speaker mean vector. Garcia-Romero and Espy-Wilson [30] found that length normalization could compensate for the non-Gaussian effect and boost performance of Gaussian PLDA to the level of the heavy-tailed PLDA. Burget, Cumani and colleagues [31, 32] proposed a pair-wised discriminative model discriminating the true speakers and the imposters. In their approach, the model accepted a pair of i-vectors and predicted the probability that how they belong to the same speaker. The input features of the model were derived from the i-vector pairs according to a form derived from the PLDA score function (further generalized to any symmetric score functions in [32]), and the model was trained on i-vector pairs that have been labelled as identical or different speakers. A particular shortcoming of this approach was that the feature expansion was highly complex. To solve this problem, a partial discriminative training approach was proposed in

[33], which optimized the discriminative model on a subspace without requiring any feature expansion. In [34], Wang proposed a discriminative approach based on deep neural networks (DNN), sharing the same idea as in the pair-wised training, whereas the features were defined manually.

2.2.3 Score Normalization

The score normalization algorithms include Hnorm [35, 36], Htnorm [37], Cnorm [38], and Atnorm [39], which utilize some a priori knowledge of channels to normalize impostors' verification scores into a standard normal distribution, so as to remove the influence of channel distortions from verification scores.

2.3 Multiple Speakers

Speaker recognition has generally been viewed as a problem of verifying or identifying a particular speaker in a speech segment containing only a single speaker. But for some real applications, the problem is to verify or identify particular speakers in a speech segment containing multiple speakers [40, 41]. Due to the diversity of multiple speaker speech, it is much more complex than a single speaker recognition and it requires high robustness to the existing system.

In a multiple-speaker scenario, if the system cannot separate single speaker segments effectively, it will directly affect the system performance. Automatic systems need to be able to segment the speech containing multiple speakers into segments and to determine whether the speech by a particular speaker is present and where in the segment this speech occurs [41], and then a series of single speaker recognition approaches could be performed. Figure 2.1 is a basic framework of a multiple-speaker recognition system. When multiple-speaker speech coming, the first procedure performs noise reduction to purify the speech audio. Following, the feature extraction and the speech activity detection is then normally performed to remove the influence of non-speech segments. Single speaker segments are extracted with speaker segmentation and clustering, and then recognition performs the same way as in single speaker recognition. Current research directions in multiple-speaker tasks include: robust features, robust speaker models, and segmentation and clustering algorithms. Robust features focus on extracting effective features in multiple-speaker scenarios, apart from MFCC, time-delay features,

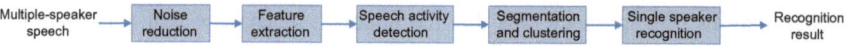

Fig. 2.1 System framework of multiple-speaker recognition

prosodic features and voice-quality features can be beneficial. Robust speaker models describe speakers in a short-time pronunciation to make more stable speaker representation. Moreover, the speaker segmentation and the clustering algorithms are two core techniques in multiple-speaker recognition. After several iterations of segmentation and clustering till convergence, the performance of multi-speaker recognition system will be improved.

2.3.1 Robust Features

One of the factors that critically affects the performance of a multiple-speaker task is the process of feature extraction. MFCC is one of the most commonly used short-term speech features in speaker recognition, and also effectively applied to multiple-speaker tasks. Besides, the time-delay feature is successful for speaker localization especially under the multiple microphone channels. Assuming that the position of any speaker does not change, the speaker localization may thus be used as alternative features in multiple-speaker tasks [42], which has become extremely popular. Some research also combined acoustic features and inter-channel delay features at the weighted log-likelihood level [43]. The use of prosodic features is emerging as a reaction to the theoretical inconsistency derived from using MFCC features for both speaker recognition and speech recognition. In [44], the authors presented a systematic investigation which showed both short-term cepstral features and long-term prosodic feature can provide significant information for speaker discrimination. References [45, 46] fused the above two features with jitter and shimmer voice-quality features and achieved considerable improvements.

2.3.2 Robust Speaker Models

One of the key points in multiple-speaker recognition is how to accurately represent speakers within a short utterance. So robust speaker modeling has become another research topic to improve the multiple-speaker recognition performance. The eigenvoice-based modeling has shown its advantages to represent speakers in speaker segmentation tasks [47], and it gets the *prior* knowledge about the speaker space to find a low dimensional vector of speaker factors that summarize the salient speaker characteristics. The speaker factors can be computed effectively in a small size window and do not suffer the problem of data sparseness. Reference [48] used an i-vector-based approach to search the speaker change points with the same idea. In order to enhance the stability and hence to improve the performance, Ref. [49] proposed a method based on a set of reference speaker models which can make a representation of the whole speaker model space. Recently, with the great success of deep neural network, Ref. [50] first proposed a novel deep neural architecture (DNN) especially for learning speaker-specific characteristics from MFCC, and used this kind of speaker-specific characteristics to perform speaker segmentation and clustering. Reference [51] compared 4 types of neural network feature transforms and found that classification networks can achieve better result than comparison networks in multiple-speaker tasks.

2.3.3   Segmentation and Clustering Algorithms

There are two types of algorithms for speaker segmentation and clustering. One is to unity the segmentation and clustering tasks into one-step, and the other is to perform the segmentation and the clustering tasks independently. The former is to identify the speaker information while getting the speaker segments [52]; The latter is to segment the audio from multiple speakers into speech segments from single speakers, and then cluster the segments from the same speakers for independent identification.

The state-of-the-art one-step segmentation and clustering algorithm is based on E-HMM models [53]. Each of them can fit into two categories: the bottom-up [54, 55] and the top-down [56] approach. The bottom-up approach is initialized with many clusters (usually more clusters than actual speakers), and the top-down approach is initialized with one cluster. In both cases, the aim is to iteratively converge towards an optimum number of clusters. Reference [57] made a comparative study of these two approaches and concluded that the bottom-up approach can capture comparatively purer models and thus can be more sensitive to nuisance variation such as the speech content; while the top-down approach can produce less discriminative speaker models but can potentially better normalized against nuisance variation. To solve the problem against initialization parameter variation in the bottom-up approach, Ref. [58] presented a method to reduce manual tuning of these values. In the E-HMM segmentation and clustering approach, the problem is rendered particularly difficulty by the fact that there is not any *a prior* knowledge of the number of speakers. Reference [59] addressed this problem with the hierarchical Dirichlet process hidden Markov model (HDP-HMM) and sticky HDP-HMM. Reference [60] proposed a process for including priors of speaker counting with agglomerative hierarchical clustering and demonstrated significantly improvement in terms of calibration error for speaker diarization.

When performing segmentation and clustering separately, some distance measures need to be pre-defined to detect speaker change points in the segmentation step. The Bayesian Information Criterion (BIC) is one kind of method for model selection and was used as the distance measure in [61]. The Generalized Likelihood Ratio (GLR) [62] is another effective distance measure method. Reference [63] used Kullback-Leibler Divergence (KL) to perform speaker segmentation in broadcast news tasks and achieved great success. Support Vector Machine (SVM) is a type of expeditious classification and was used for segmentation in [64]. In the clustering step, in order to reduce the cost of computation, the agglomerative information bottleneck (aIB) [65] and the sequential information bottleneck (sIB) [66] were proposed and they therefore were widely used in the meeting scenarios. Reference [67] made a noise-robust speaker clustering based on spectral clustering and compared it with the hierarchical clustering and the K-means clustering. Reference [68] proposed a novel DNN-based clustering and performed re-segmentation and clustering in each iteration. Reference [69] investigated how accurate the clustering algorithm will be depending on the characteristics of the audio stream, which was an effective guidance of speaker clustering. ELISA [70] was a hybrid system combining the segmentation and the clustering steps together.

2.4 Discussions

Background noise, channel mismatch and multiple speakers are three most common factors that will influence the performance of speaker recognition systems. In real applications, there is often *no prior* knowledge of environmental noise, transmission channel and number of speakers containing in the speech segment in advance. Therefore, it is difficult to pre-train the noise/channel model and define the clustering number. To deal with these environment-related issues, researchers have carried to do some studies from different of views. In this chapter, we summarize the latest research studies and techniques among these three factors from different aspects. We believe that these three factors are still the main research directions.

References

1. Boll S (1979) Suppression of acoustic noise in speech using spectral subtraction. IEEE Trans Acoust Speech Signal Process 27(2):113–120
2. Wang N, Ching PC, Zheng N et al (2011) Robust speaker recognition using denoised vocal source and vocal tract features. IEEE Trans Audio Speech Lang Process 19(1):196–205
3. Hermansky H, Morgan N (1994) RASTA processing of speech. IEEE Trans Speech Audio Process 2(4):578–589
4. Furui S (1981) Cepstral analysis technique for automatic speaker verification. IEEE Trans Acoust Speech Signal Process 29(2):254–272
5. Zhao X, Shao Y, Wang DL (2012) CASA-based robust speaker identification. IEEE Trans Audio Speech Lang Process 20(5):1608–1616
6. Sadjadi SO, Hansen JHL (2010) Assessment of single-channel speech enhancement techniques for speaker identification under mismatched conditions. INTERSPEECH, pp 2138–2141
7. Hanilçi C, Kinnunen T, Saeidi R et al (2012) Comparing spectrum estimators in speaker verification under additive noise degradation. Acoustics, speech and signal processing (ICASSP), 2012 IEEE international conference on. IEEE, pp 4769–4772
8. Lei Y, Burget L, Scheffer N (2013) A noise robust i-vector extractor using vector taylor series for speaker recognition. Acoustics, speech and signal processing (ICASSP), 2013 IEEE international conference on. IEEE, pp 6788–6791

9. Lei Y, McLaren M, Ferrer L et al (2014) Simplified vts-based i-vector extraction in noise-robust speaker recognition. Acoustics, speech and signal processing (ICASSP), 2014 IEEE international conference on. IEEE, pp 4037–4041

10. Martinez D, Burget L, Stafylakis T et al (2014) Unscented transform for ivector-based noisy speaker recognition. Acoustics, speech and signal processing (ICASSP), 2014 IEEE international conference on. IEEE, pp 4042–4046

11. Gales MJF, Young SJ (1996) Robust continuous speech recognition using parallel model combination. IEEE Trans Speech Audio Process 4(5):352–359

12. Bellot O, Matrouf D, Merlin T et al (2000) Additive and convolutional noises compensation for speaker recognition. INTERSPEECH, pp 799–802

13. Lei Y, Burget L, Ferrer L et al (2012) Towards noise-robust speaker recognition using probabilistic linear discriminant analysis. Acoustics, speech and signal processing (ICASSP), 2012 IEEE international conference on. IEEE, pp 4253–4256

14. Doddington GR, Przybocki MA, Martin AF et al (2000) The NIST speaker recognition evaluation–overview, methodology, systems, results, perspective. Speech Commun 31 (2):225–254

15. The NIST year 2012 speaker recognition evaluation plan. https://www.nist.gov/sites/default/files/documents/itl/iad/mig/NIST_SRE12_evalplan-v17-r1.pdf

16. NIST 2016 speaker recognition evaluation plan. https://www.nist.gov/sites/default/files/documents/itl/iad/mig/SRE16_Eval_Plan_V1-0.pdf

17. Furui S (1981) Cepstral analysis technique for automatic speaker verification. IEEE Trans Acoust Speech Signal Process 29(2):254–272

18. Reynolds DA (2003) Channel robust speaker verification via feature mapping. Acoustics, speech, and signal processing, 2003. Proceedings. (ICASSP'03). 2003 IEEE international conference on. IEEE, vol 2, pp 2–53

19. Teunen R, Shahshahani B, Heck LP (2000) A model-based transformational approach to robust speaker recognition. INTERSPEECH, pp 495–498

20. Wu W, Zheng TF, Xu MX et al (2007) A cohort-based speaker model synthesis for mismatched channels in speaker verification. IEEE Trans Audio Speech Lang Process 15 (6):1893–1903

21. Solomonoff A, Quillen C, Campbell WM (2004) Channel compensation for SVM speaker recognition. Odyssey, vol 4, pp 219–226

22. Matrouf D, Scheffer N, Fauve BGB et al (2007) A straightforward and efficient implementation of the factor analysis model for speaker verification. INTERSPEECH, pp 1242–1245

23. Kenny P, Boulianne G, Ouellet P et al (2007) Joint factor analysis versus eigenchannels in speaker recognition. IEEE Trans Audio Speech Lang Process 15(4):1435–1447

24. Dehak N, Kenny PJ, Dehak R et al (2011) Front-end factor analysis for speaker verification. IEEE Trans Audio Speech Lang Process 19(4):788–798

25. Hatch AO, Kajarekar SS, Stolcke A (2006) Within-class covariance normalization for SVM-based speaker recognition. INTERSPEECH

26. McLaren M, Van Leeuwen D (2011) Source-normalised-and-weighted LDA for robust speaker recognition using i-vectors. Acoustics, speech and signal processing (ICASSP), 2011 IEEE international conference on. IEEE, pp 5456–5459

27. Ioffe S (2006) Probabilistic linear discriminant analysis. European conference on computer vision. Springer, Berlin, pp 531–542

28. Prince SJD, Elder JH (2007) Probabilistic linear discriminant analysis for inferences about identity. Computer vision, 2007. ICCV 2007. IEEE 11th international conference on. IEEE, pp 1–8

29. Kenny P (2010) Bayesian speaker verification with heavy-tailed priors. Odyssey, pp 14

30. Garcia-Romero D, Espy-Wilson CY (2011) Analysis of i-vector length normalization in speaker recognition systems. INTERSPEECH, pp 249–252

31. Burget L, Plchot O, Cumani S et al (2011) Discriminatively trained probabilistic linear discriminant analysis for speaker verification. Acoustics, speech and signal processing (ICASSP), 2011 IEEE international conference on. IEEE, pp 4832–4835

32. Cumani S, Brummer N, Burget L et al (2013) Pairwise discriminative speaker verification in the i-vector space. IEEE Trans Audio Speech Lang Process 21(6):1217–1227

33. Hirano I, Lee KA, Zhang Z et al (2014) Single-sided approach to discriminative PLDA training for text-independent speaker verification without using expanded i-vector. Chinese spoken language processing (ISCSLP), 2014 9th international symposium on. IEEE, pp 59–63

34. Wang J, Wang D, Zhu Z et al (2014) Discriminative scoring for speaker recognition based on i-vectors. Asia-pacific signal and information processing association, 2014 annual summit and conference (APSIPA). IEEE, pp 1–5

35. Reynolds DA, Quatieri TF, Dunn RB (2000) Speaker verification using adapted Gaussian mixture models. Digit Signal Proc 10(1–3):19–41

36. Reynolds DA (1997) Comparison of background normalization methods for text-independent speaker verification. Eurospeech

37. Auckenthaler R, Carey M, Lloyd-Thomas H (2000) Score normalization for text-independent speaker verification systems. Digit Signal Proc 10(1):42–54

38. Bimbot F, Bonastre JF, Fredouille C et al (2004) A tutorial on text-independent speaker verification. EURASIP J Appl Sig Process 2004:430–451

39. Sturim DE, Reynolds DA (2005) Speaker adaptive cohort selection for Tnorm in text-independent speaker verification. Acoustics, speech, and signal processing, 2005. Proceedings. (ICASSP'05). IEEE international conference on. IEEE, 1: I/741-I/744 vol 1

40. Anguera X, Bozonnet S, Evans N et al (2012) Speaker diarization: a review of recent research. IEEE Trans Audio Speech Lang Process 20(2):356–370

41. Martin AF, Przybocki MA (2001) Speaker recognition in a multi-speaker environment. INTERSPEECH, pp 787–790

42. Lathoud G, McCowan IA (2003) Location based speaker segmentation. Multimedia and expo, 2003. ICME'03. Proceedings. 2003 international conference on. IEEE, vol 3, pp 3–621

43. Pardo JM, Anguera X, Wooters C Speaker diarization for multiple distant microphone meetings: mixing acoustic features and inter-channel time differences. INTERSPEECH

44. Friedland G, Vinyals O, Huang Y et al (2009) Prosodic and other long-term features for speaker diarization. IEEE Trans Audio Speech Lang Process 17(5):985–993

45. Woubie A, Luque J, Hernando J (2015) Using voice-quality measurements with prosodic and spectral features for speaker diarization. INTERSPEECH, pp 3100–3104

46. Woubie A, Luque J, Hernando J (2016) Short-and long-term speech features for hybrid HMM-i-Vector based speaker diarization system. Odyssey

47. Castaldo F, Colibro D, Dalmasso E et al (2008) Stream-based speaker segmentation using speaker factors and eigenvoice. Acoustics, speech and signal processing, 2008. ICASSP 2008. IEEE international conference on. IEEE, pp 4133–4136

48. Desplanques B, Demuynck K, Martens JP (2015) Factor analysis for speaker segmentation and improved speaker diarization. INTERSPEECH. Abstracts and proceedings USB productions, pp 3081–3085

49. Wang G, Zheng TF (2009) Speaker segmentation based on between-window correlation over speakers' characteristics. Proceedings: APSIPA ASC, pp 817–820

50. Chen K, Salman A (2011) Learning speaker-specific characteristics with a deep neural architecture. IEEE Trans Neural Networks 22(11):1744–1756

51. Yella SH, Stolcke A (2015) A comparison of neural network feature transforms for speaker diarization. INTERSPEECH, pp 3026–3030

52. Kotti M, Moschou V, Kotropoulos C (2008) Speaker segmentation and clustering. Sig Process 88(5):1091–1124

53. Meignier S, Bonastre JF, Igounet S (2001) E-HMM approach for learning and adapting sound models for speaker indexing. A speaker Odyssey-the speaker recognition workshop

54. Meignier S, Bonastre JF, Fredouille C et al (2000) Evolutive HMM for multi-speaker tracking system. Acoustics, speech, and signal processing, 2000. ICASSP'00. Proceedings. 2000 IEEE international conference on. IEEE, vol 2, pp 1201–1204
55. Ajmera J, Wooters C (2003) A robust speaker clustering algorithm. Automatic speech recognition and understanding, 2003. ASRU'03. 2003 IEEE Workshop on. IEEE, pp 411–416
56. Wooters C, Huijbregts M (2008) The ICSI RT07s speaker diarization system. Multimodal technologies for perception of humans. Springer, Berlin, pp 509–519
57. Evans N, Bozonnet S, Wang D et al (2012) A comparative study of bottom-up and top-down approaches to speaker diarization. IEEE Trans Audio Speech Lang Process 20(2):382–392
58. Imseng D, Friedland G (2010) Tuning-robust initialization methods for speaker diarization. IEEE Trans Audio Speech Lang Process 18(8):2028–2037
59. Fox EB, Sudderth EB, Jordan MI et al (2011) A sticky HDP-HMM with application to speaker diarization. The annals of applied statistics, pp 1020–1056
60. Sell G, McCree A, Garcia-Romero D (2016) Priors for speaker counting and diarization with AHC. INTERSPEECH 2016, pp 2194–2198
61. Chen S, Gopalakrishnan P (1998) Speaker, environment and channel change detection and clustering via the bayesian information criterion. Proc. DARPA broadcast news transcription and understanding workshop, vol 8, pp 127–132
62. Gish H, Siu MH, Rohlicek R (1991) Segregation of speakers for speech recognition and speaker identification. Acoustics, speech, and signal processing, 1991. ICASSP-91, 1991 international conference on. IEEE, pp 873–876
63. Siegler MA, Jain U, Raj B et al (1997) Automatic segmentation, classification and clustering of broadcast news audio. Proc. DARPA speech recognition workshop. 1997
64. Fergani B, Davy M, Houacine A (2008) Speaker diarization using one-class support vector machines. Speech Commun 50(5):355–365
65. Vijayasenan D, Valente F, Bourlard H (2007) Agglomerative information bottleneck for speaker diarization of meetings data. Automatic speech recognition and understanding, 2007. ASRU. IEEE workshop on. IEEE, pp 250–255
66. Vijayasenan D, Valente F, Bourlard H (2008) Combination of agglomerative and sequential clustering for speaker diarization. Acoustics, speech and signal processing, 2008. ICASSP 2008. IEEE international conference on. IEEE, pp 4361–4364
67. Tawara N, Ogawa T, Kobayashi T (2015) A comparative study of spectral clustering for i-vector-based speaker clustering under noisy conditions. Acoustics, speech and signal processing (ICASSP), 2015 IEEE international conference on. IEEE, pp 2041–2045
68. Milner R, Hain T (2016) DNN-based speaker clustering for speaker diarisation. Proceedings of the annual conference of the international speech communication association, INTERSPEECH. Sheffield, pp 2185–2189
69. Prieto JJ, Vaquero C, García P (2016) Analysis of the impact of the audio database characteristics in the accuracy of a speaker clustering system. Odyssey, pp 393–399
70. Moraru D, Meignier S, Fredouille C et al (2004) The ELISA consortium approaches in broadcast news speaker segmentation during the NIST 2003 rich transcription evaluation. Acoustics, speech, and signal processing, 2004. Proceedings. (ICASSP'04). IEEE international conference on. IEEE, vol 1, pp 1–373

Chapter 3
Speaker-Related Robustness Issues

Speaker dependent factors, such as gender, physical condition (cold or laryngitis), speaking style (emotion state, speech rate, etc.), cross-language, accent and session variations, are major concerns in speech signal processing. How they correlate with each other and what the key factors are in speech realization are real considerations in research [1]. The current mainstream research can be divided into five directions which will be described in the following subsections.

3.1 Genders

A speaker recognition system can easily achieve its best accuracy when trained with gender dependent (GD) data and tested with *priori* gender information. However, in real applications, gender information is often not available. Therefore, how to design a system that does not rely on the gender labels has highly practical significance. Reference [2] addressed the problem of designing a fully gender independent (GI) speaker recognition system. It relied on discriminative training, where the trails were i-vector pairs, and the discrimination was between the hypothesis that the pair of feature vectors in the trial belong to the same speaker or different speakers. They demonstrated that this pairwise discriminative training could be interpreted as a procedure that estimates the parameters of the best (second order) approximation of the log-likelihood ratio score function, and a pairwise SVM could be used for training a GI system. The results showed that a fully GI system had been trained which was slightly worse than a fully GD systems.

A novel GI PLDA classifier for i-vector based speaker recognition was proposed and evaluated in [3]. This approach was to use the source-normalization for variation that separates genders as a pre-processing step for i-vector based PLDA classification. Experimental results on the NIST 2010 SRE dataset demonstrated that it could reach comparable performance compared to a typical GD configuration.

© The Author(s) 2017
T.F. Zheng and L. Li, *Robustness-Related Issues in Speaker Recognition*,
SpringerBriefs in Signal Processing, DOI 10.1007/978-981-10-3238-7_3

3.2   Physical Conditions

Speech is a behavioral signal that may not be consistently reproduced by a speaker and can be affected by a speaker's physical conditions. This variability aspect may be due to illness such as cold, nasal congestion, laryngitis and other behavioral variations. As early as in 1996, Tull and his team had carried out related research. They studied on the "cold-affected" speech in speaker recognition [4, 5]. They analyzed the differences between "cold-affected" speech and normal/healthy speech from the resonances of the vocal tract, fundamental frequency (measurements of voice pitch), phonetic differences and Mel-cepstral coefficients. And they found that "cold-speech" shows some noisy portions in the acoustic signal that are not present in the normal/healthy signals and these noisy portions are caused by hoarseness and coughing. Besides, they also analyzed phonetic contrasts and looked at differences in formant patterns. Phonetic transcriptions of cold and normal sessions reflect changes in place of articulation. Perceptual and acoustic analyses revealed that pauses and epenthetic syllables are not constant throughout all sessions. These differences suggested that the cold introduces another level of intra-speaker variability that needs to be addressed in the design of speaker recognition systems.

Nowadays, research in this direction is still rare and most researchers are just focusing in the feature level [6, 7]. The reason is perhaps that this type of speech data is difficult to collect. But in real situations, a person's physical condition is variable and will have great effect on the performance of speaker recognition. Therefore, research on robustness to the speaker-physical condition variability has very practical importance.

3.3   Speaking Styles

The speaking style usually varies when a person speaks in a spontaneous way and it contains the person's abundant individual information. For speech recognition, there has been a lot of research on the speaking style [8–10]. The major methods are to choose more emotion-related features and train multi-style speaking models. In recent years, many researchers attempted to apply these methods in speaker recognition systems. In this section, we will review the speaking style analysis of speaker recognition in three aspects.

3.3.1   Emotion

Emotion is an intrinsic nature of human beings and changes in the rendering forms of speech signals significantly. Compared with other intra-speaker variations such as the speaking rate and speech tones, emotions tend to cause more substantial variations on speech properties, such as harmonic forms, formant structures and the entire temporal-spectral patterns [11].

A multitude of research has been conducted to address the emotion variations. The first category involves analysis of various emotion-related acoustic factors such as prosody and voice quality [12], pitch [13, 14], duration and sound intensity [14]. The second category involves various emotion-compensation methods for modeling and scoring. An early investigation was supported by the European VERIVOX project [15, 16], where the researchers proposed a 'structured training' which elicited the enrollment speech in various speaking styles. By training the speaker models with the elicited multi-emotional speech, the authors reported reasonable performance improvements. This method, however, had poor interactivity and was unacceptable in practice. Wu et al. [13] presented an emotion-added model training method, where a few emotional data were used to train emotion-dependent models. Besides, [17] compared three types of speaker models (HMM, circular HMM, and supra-segmental HMM) and concluded that the supra-segmental HMM was most robust against emotion changes. In addition, [18] proposed an adaption approach based on the maximum likelihood linear regression (MLLR) and its feature-space variant, the constrained MLLR (CMLLR). The basic idea here was to project the emotional test utterances to neutral utterances by a linear transformation. Then they presented a novel emotion adaptive training (EAT) method to iteratively estimate the emotion-dependent CMLLR transformations and re-train the speaker models with transformed speech. The results demonstrated that the EAT approach achieved significant performance improvements over the baseline system where the neutral enrollment data was used to train the speaker models and the emotional test utterances were tested directly. Score normalization and transformation approaches [19, 20] have also been proposed to improve emotional speaker recognition.

3.3.2   Speaking Rate

The speaking rate is another high-level speaker variable. It can be useful for discriminating between speakers. Speech events in two different utterances-even though they have the same text and are spoken by the same speaker-are seldom synchronized in time. This effect is attributable to the differences in the speaking rates [21]. The speaking rate is often observed in practical speaker recognition systems. For instance, people tend to speak faster if he/she is in rush. Conversely, one may speak slowly due to the exhaustion or illness. A small difference in speaking rate between the enrollment and the test speech will not be a problem, but substantial mismatch may lead to serious performance degradation.

The research on the speaking rate is still preliminary in speaker recognition. Early studies mainly focused on the impact of speaking rates to speaker recognition. For example, performance degradation was confirmed by [22] when the mismatch on speaking rates (fast and slow) exists. For text-dependent speaker recognition, the non-linear time alignment such as DTW (Dynamic Time Warping) algorithm was also used to solve the speaking rate variation problem. A normalization approach was proposed to mitigate the impact of speaking rate mismatch [23]. In this approach, the phoneme duration was used as an additional feature and was augmented to the conventional Mel frequency cepstral coefficients (MFCCs). The experiments on the YOHO corpus confirmed that with this normalization, both robustness and accuracy of speaker verification could be improved. Reference [22] proposed a feature transform approach which projected speech features in slow speaking rates to those in the normal speaking rate. The feature space maximum likelihood linear regression (fMLLR) was adopted to conduct the transform, under the well-known GMM-UBM and i-vector framework. Experiments showed that the proposed fMLLR-based approach could diminish this reduction when there was a mismatch in the speaking rate between enrollment and test.

3.3.3 Idiom

Idiom, a person's personal style of word usage, is a high-level intra-speaker characteristic, and can discriminate speakers coming from different regions or having different educational backgrounds. For instance, we humans can recognize a familiar speaker by only a couple of idioms from him/her. That means through self-learning and training, the human brain can perform speaker recognition with the aid of idioms. Scholars in the field of linguistics opened the studies on idiom, including idiom principles [24], idiom structure [25], idiomatic expressions [26], idiom tagging [27], and etc. There have been a lot of investigations for automatic speaker recognition with high-level idiom information, such as idiosyncratic word-usage high-level feature [28–30] and idiosyncratic pronunciation feature [31–34].

Actually, idiom is not an adverse factor for robust speaker recognition; on the contrary, it can be used as a high-level feature to improve robustness. We list it here because if not well dealt with, as a kind of speaking style factor, it may affect the recognition performance.

3.4 Cross Languages

The language is a fundamental capability of human beings. In history, the language in a particular area was fairly stable in most of the time because of the limited inter-area or inter-nation cultural and business exchanging. With the rapid development of international business such as bilateral trading, this stability has been fundamentally changed since last century. Especially in the latest decades, the development of the Internet has been leading to much more efficient information exchange and has glued the entire world together. An interesting consequence of this change in human language is the bilingual or multilingual phenomena, leading to the cross-language issue. This issue occurs when the enrollment is in one language and the test is in another. For instance, many people can speak both English and Chinese, and it is very likely that one enrolls in English and tests in Chinese, or vice versa, when he/she uses speaker recognition systems. This mismatch often leads to serious performance degradation. Considering the large population of bilingual speakers, the research on bilingual and cross-lingual speaker verification is desirable for a wide range of applications.

The bilingual and cross-lingual challenges can be largely attributed to the different distributions of acoustic features in different languages. Specifically, most of the current speaker verification systems rely on statistical models that are adopted in this work. A basic assumption of the statistical model is that the distributions of acoustic features are identical in enrollment and in test. This is certainly not true if the enrollment and test speech are in different languages.

The research on bilingual and cross-lingual speaker verification is still in the preliminary stage. The early studies investigated the impact of language mismatch. For example in [35], the authors investigated language mismatch between UBM training and speaker enrollment. Their experiments confirmed a considerable performance degradations if the UBM was trained in one language and the speaker models were trained in another. If the UBM was trained with data pooled from multiple languages including the one used in speaker enrollment/test, the performance degradation was mitigated. A similar study was proposed in [36], and the focus in that paper was the language mismatch between speaker enrollment and test. The authors demonstrated that this mismatch led to significant performance degradation. Again, if the enrollment data involved multiple languages including the one used in test, the performance degradation could be largely mitigated. In [37], applying the t-SNE method [38, 39], authors draw the mean vectors of Gaussian components as points in a two-dimensional space. The difference between the Standard Chinese and Uyghur models can be recognized from the distributions of the Gaussian mean vectors in this space, as shown in Fig. 3.1. It is observed that the Gaussian components of the two languages clearly deviate from each other. This means that if a speaker model is enrolled in one language, it would be difficult for the model to represent speech signals in another language.

Another approach to the bilingual and cross-lingual challenges is to employ language-independent acoustic features, such as the residual phase cepstral coefficients (RPCCs) and the glottal glow cepstral coefficients (GLFCCs) [40]. These features were demonstrated to be less dependent on phonetic content of speech signals, therefore lending themselves to tasks with language mismatch. Another advantage of the RPCC and the GLFCC features is that they may deliver better performance if the training data is limited, which is often the case in cross-lingual scenarios. Besides, [41] proposed two novel algorithms towards the spoken language mismatch problem: the first algorithm merged the language-dependent system outputs by using the Language Identification (LID) scores of each utterance as the fusion weights. In the second algorithm, fusion was performed at the segment level via multilingual Phone Recognition (PR). In [42], the authors proposed a

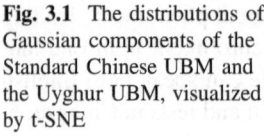 **Fig. 3.1** The distributions of Gaussian components of the Standard Chinese UBM and the Uyghur UBM, visualized by t-SNE

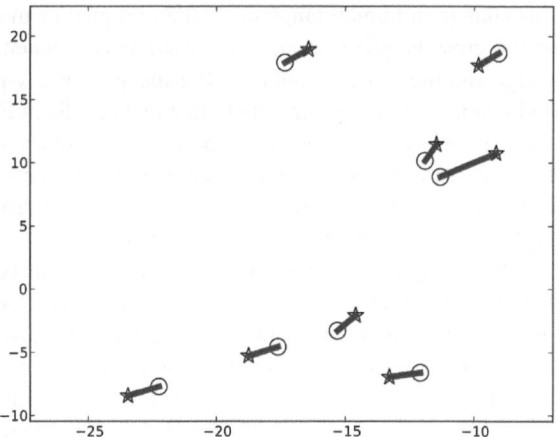

method of combining features (MFCC and LPCC) to obtain a more robust multilingual speaker recognition system. The experimental results indicated that this method could be used to improve the speaker identification performance in multilingual conditions with limited data.

The third category of approach to the cross-lingual problem contains various factor analysis methods. These methods assume that the enrollment data of a speaker covers all the languages encountered in test, however the language in a particular trial is unknown. Factor analysis treats the language as a latent variable and marginalizes it out when conducting verification. For example, the method proposed in [43] inferred the posterior probabilities of a test speech belonging to different languages. With these posteriors, the score of the test speech against the claimed speaker was computed as a linear fusion of the scores tested against the speaker models enrolled with each language. The utterance-based fusion was further extended to the segment-based fusion, where the posteriors probabilities over languages were computed for each phone segment and then the segment-level speaker scores were fused according to these posteriors. The work in [44] followed in the same direction but formulated the problem in a more elegant Bayesian framework. In that work, the joint factor analysis (JFA) formulation was extended by adding a latent factor to represent the language. This language factor was inferred and compensated for enrollment and test.

3.5   Time Varying

It is believed in the speaker recognition field that there exists identifiable uniqueness in each voice. At the same time, a question arises: whether the voice changes significantly with time [45]. Similar ideas were expressed in [46, 47], as they argued that a big challenge to uniquely characterize a person's voice was that voice changes over time. Performance degradation has also been observed in the presence of time intervals in a practical speaker recognition system [48–50]. From the pattern recognition point of view, the enrollment data (for training speaker models) and the test utterances for verification are usually separated by some period of time, which poses a mismatch leading to recognition performance degradation.

As shown in Fig. 3.2, the EER curves can demonstrate how the performance of the speaker verification system changes with time elapse as shown by the black line with solid dots. Here, all sessions are distributed along the horizontal axis according to their time intervals from the second session (day 0) in days.

Therefore, some researchers resorted to several training sessions over a long period of time for dealing with long-term variability issues [51]. In [52], the best recognition performance was achieved when 5 sessions successively separated by at least 1 week were used to form the training set. References [53, 54] used a similar method called data augmentation. In this approach, when a positive identification of the candidate speaker was valid, extra data was appended to the original enrollment data to provide a more universal enrollment model for this candidate. This approach requires the original data to be stored for re-enrollment. At the same time, speaker adaptation techniques, such as MAP-adaptation [53, 54] and MLLR-adaptation [55], were proposed to adapt the original model to a new model with new data so as to reduce the effects of the ageing issue. In [55], experimental results showed that by adapting the speaker models using data from the intervening sessions, the Equal Error Rate (EER) on the last two sessions was reduced from 2.5 to 1.7%.

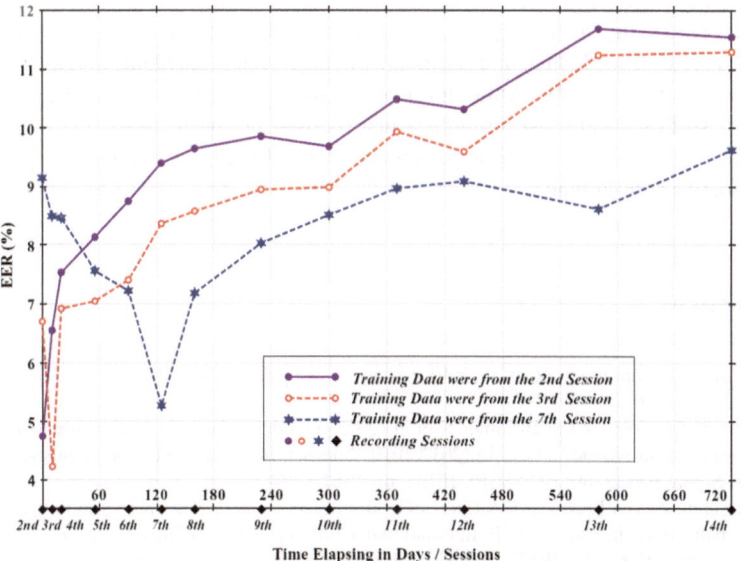

Fig. 3.2 Three *curves* of EERs corresponding to three different training sessions [58]

In the score domain, researchers observed that the verification scores of true speakers decreased progressively as the time span between enrollment and verification increased, while the scores of impostors were less affected [56, 57]. Thus, a stacked classifier method with an ageing-dependent decision boundary was applied to improve the long-term verification accuracy.

From a different direction, Wang et al. [58] aimed to extract and combine the speaker-sensitive and the time-insensitive information as features to deal with this long-term variability. Wang investigated the issues in the frequency domain by emphasizing the discrimination of information with higher speaker-specific sensitivity and lower time-related session-specific sensitivity. F-ratio was employed as a criterion to determine the figure of merit to judge the foresaid two kinds of sensitivities, and to find a compromise between them. Inspired by the feature extraction procedure of the traditional MFCC calculation, two emphasis strategies were explored when generating modified acoustic features, the pre-filtering frequency warping and the post-filtering filter-bank outputs weighting were used for speaker verification. On a time-varying voiceprint database [59], experimental results showed that the proposed features outperformed both MFCCs and LPCCs, and to some extent, alleviated the ageing impact on speaker recognition.

3.6 Discussion

In real life situations, the pronunciation characteristics of a person is polytropic and uncontrollable. So how to build a robust speaker recognition system against these speaker-related factors is of great significant. In this chapter, five speaker-related factors are discussed, they are genders, physical state, speaking style, cross languages, and time varying. Although it seems that research on these speaker-related issues is still rare and most researches just focused on the feature level, we believe that research on robustness to these speaker-related factors has good prospects and deep practical importance.

References

1. Huang C, Chen T, Li SZ et al (2001) Analysis of speaker variability. In: INTERSPEECH. pp 1377–1380
2. Cumani S, Glembek O, Brümmer N et al (2012) Gender independent discriminative speaker recognition in i-vector space. In: 2012 IEEE International Conference on Acoustics, Speech and Signal Processing (ICASSP). IEEE, pp 4361–4364
3. McLaren M, van Leeuwen DA (2012) Gender-independent speaker recognition using source normalization. In: 2012 IEEE International Conference on Acoustics, Speech and Signal Processing (ICASSP). IEEE, pp 4373–4376
4. Tull RG, Rutledge JC (1996) Analysis of "cold-affected" speech for inclusion in speaker recognition systems. J Acoust Soc Am 99(4):2549–2574

5. Tull RG, Rutledge JC (1996) 'Cold speech' for automatic speaker recognition. In: Acoustical Society of America 131st Meeting Lay Language Papers
6. Tull RG, Rutledge JC, Larson CR (1996) Cepstral analysis of "cold-speech" for speaker recognition: a second look. Dissertation, ASA
7. Tull RG (1999) Acoustic analysis of cold-speech: implications for speaker recognition technology and the common cold. Northwestern University
8. Kwon OW, Chan K, Hao J et al (2003) Emotion recognition by speech signals. In: INTERSPEECH
9. Juang BH (1991) Speech recognition in adverse environments. Comput Speech Lang 5 (3):275–294
10. Lippmann R, Martin E, Paul D (1987) Multi-style training for robust isolated-word speech recognition. In: IEEE International Conference on Acoustics, Speech, and Signal Processing, ICASSP'87, vol 12. IEEE, pp 705–708
11. Bie F, Wang D, Zheng TF et al (2013) Emotional speaker verification with linear adaptation. In: 2013 IEEE China Summit & International Conference on Signal and Information Processing (ChinaSIP). IEEE, pp 91–94
12. Zetterholm E (1998) Prosody and voice quality in the expression of emotions. In: ICSLP
13. Wu T, Yang Y, Wu Z (2005) Improving speaker recognition by training on emotion-added models. In: International Conference on Affective Computing and Intelligent Interaction. Springer, Berlin, Heidelberg, pp 382–389
14. Pereira C, Watson CI (1998) Some acoustic characteristics of emotion. In: ICSLP
15. Scherer KR, Johnstone T, Klasmeyer G et al (2000) Can automatic speaker verification be improved by training the algorithms on emotional speech? In: INTERSPEECH. pp 807–810
16. Scherer KR, Grandjean D, Johnstone T et al Acoustic correlates of task load and stress. In: INTERSPEECH
17. Shahin I (2009) Speaker identification in emotional environments. Iran J Electr Comput Eng 8 (1):41–46
18. Bie F, Wang D, Zheng TF et al (2013) Emotional adaptive training for speaker verification. In: 2013 Asia-Pacific Signal and Information Processing Association Annual Summit and Conference (APSIPA). IEEE, pp 1–4
19. Wu W, Zheng TF, Xu MX et al (2006) Study on speaker verification on emotional speech. In: INTERSPEECH
20. Shan Z, Yang Y (2008) Learning polynomial function based neutral-emotion GMM transformation for emotional speaker recognition. In: 19th International Conference on Pattern Recognition, 2008, ICPR 2008. IEEE, pp 1–4
21. Atal BS (1976) Automatic recognition of speakers from their voices. Proc IEEE 64(4):460–475
22. Rozi A, Li L, Wang D et al (2016) Feature transformation for speaker verification under speaking rate mismatch condition. In: 2016 Asia-Pacific Signal and Information Processing Association Annual Summit and Conference (APSIPA). IEEE, pp 1–4
23. van Heerden CJ, Barnard E (2007) Speech rate normalization used to improve speaker verification. In: Proceedings of the Symposium of the Pattern Recognition Association of South Africa. pp 2–7
24. Erman B, Warren B (2000) The idiom principle and the open choice principle. Text-Interdisc J Study Discourse 20(1):29–62
25. Makkai A (1972) Idiom structure in English. Walter de Gruyter
26. Cacciari C, Glucksberg S (1991) Understanding idiomatic expressions: the contribution of word meanings. Adv Psychol 77:217–240
27. Leech G, Garside R, Bryant M (1994) CLAWS4: the tagging of the British National Corpus. In: Proceedings of the 15th conference on Computational linguistics, vol 1. Association for Computational Linguistics, pp 622–628
28. Doddington GR (2001) Speaker recognition based on idiolectal differences between speakers. In: INTERSPEECH. pp 2521–2524

29. Kajarekar SS, Ferrer L, Shriberg E et al (2005) SRI's 2004 NIST speaker recognition evaluation system. In: Proceedings of the IEEE International Conference on Acoustics, Speech, and Signal Processing, 2005 (ICASSP'05), vol 1. IEEE, I/173–I/176
30. Stolcke GTESA, Kajarekar S (2007) Duration and pronunciation conditioned lexical modeling for speaker verification
31. Andrews WD, Kohler MA, Campbell JP et al (2002) Gender-dependent phonetic refraction for speaker recognition. In: 2002 IEEE International Conference on Acoustics, Speech, and Signal Processing (ICASSP), vol 1. IEEE, pp I-149–I-152
32. Navrátil J, Jin Q, Andrews WD et al (2003) Phonetic speaker recognition using maximum-likelihood binary-decision tree models. In: Proceedings of the 2003 IEEE International Conference on Acoustics, Speech, and Signal Processing, 2003 (ICASSP'03), vol 4. IEEE, p IV-796
33. Jin Q, Navratil J, Reynolds DA et al (2003) Combining cross-stream and time dimensions in phonetic speaker recognition. In: Proceedings of the 2003 IEEE International Conference on Acoustics, Speech, and Signal Processing, 2003 (ICASSP'03). IEEE, p IV-800
34. Hatch AO, Peskin B, Stolcke A (2005) Improved phonetic speaker recognition using lattice decoding. In: Proceedings of the IEEE International Conference on Acoustics, Speech, and Signal Processing, 2005 (ICASSP'05), vol 1. IEEE, pp I/169–I/172
35. Auckenthaler R, Carey MJ, Mason JSD (2001) Language dependency in text-independent speaker verification. In: Proceedings of the 2001 IEEE International Conference on Acoustics, Speech, and Signal Processing, 2001 (ICASSP'01), vol 1. IEEE, pp 441–444
36. Ma B, Meng H (2004) English-Chinese bilingual text-independent speaker verification. In: Proceedings of the IEEE International Conference on Acoustics, Speech, and Signal Processing, 2004 (ICASSP'04), vol 5. IEEE, p V-293
37. Askar R, Wang D, Bie F et al (2015) Cross-lingual speaker verification based on linear transform. In: 2015 IEEE China Summit and International Conference on Signal and Information Processing (ChinaSIP). IEEE, pp 519–523
38. Maaten L, Hinton G (2008) Visualizing data using t-SNE. J Mach Learn Res 9(Nov):2579–2605
39. Van Der Maaten L (2014) Accelerating t-SNE using tree-based algorithms. J Mach Learn Res 15(1):3221–3245
40. Wang J, Johnson MT (2013) Vocal source features for bilingual speaker identification. In: 2013 IEEE China Summit & International Conference on Signal and Information Processing (ChinaSIP). IEEE, pp 170–173
41. Akbacak M, Hansen JHL (2007) Language normalization for bilingual speaker recognition systems. In: IEEE International Conference on Acoustics, Speech and Signal Processing, 2007, ICASSP 2007, vol 4. IEEE, pp IV-257–IV-260
42. Nagaraja BG, Jayanna HS (2013) Combination of features for multilingual speaker identification with the constraint of limited data. Int J Comput Appl 70(6)
43. Akbacak M, Hansen JHL (2007) Language normalization for bilingual speaker recognition systems. In: IEEE International Conference on Acoustics, Speech and Signal Processing, 2007, ICASSP 2007, vol 4. IEEE, pp IV-257–IV-260
44. Lu L, Dong Y, Zhao X et al (2009) The effect of language factors for robust speaker recognition. In: IEEE International Conference on Acoustics, Speech and Signal Processing, 2009, ICASSP 2009. IEEE, pp 4217–4220
45. Kersta LG (1962) Voiceprint identification. J Acoust Soc Am 34(5):725
46. Furui S (1997) Recent advances in speaker recognition. Pattern Recogn Lett 18(9):859–872
47. Bonastre JF, Bimbot F, Boë LJ et al (2003) Person authentication by voice: a need for caution. In: INTERSPEECH
48. Mishra P (2012) A vector quantization approach to speaker recognition. In: Proceedings of the International Conference on Innovation & Research in Technology for sustainable development (ICIRT 2012), vol 1. p 152

49. Kato T, Shimizu T (2003) Improved speaker, verification over the cellular phone network using phoneme-balanced and digit-sequence-preserving connected digit patterns. In: Proceedings of the 2003 IEEE International Conference on Acoustics, Speech, and Signal Processing, 2003 (ICASSP'03), vol 3. IEEE, p II-57
50. Hébert M (2008) Text-dependent speaker recognition. In: Springer handbook of speech processing. Springer, Berlin, Heidelberg, pp 743–762
51. Bimbot F, Bonastre JF, Fredouille C et al (2004) A tutorial on text-independent speaker verification. EURASIP J Appl Signal Process, 430–451
52. Markel J, Davis S (1979) Text-independent speaker recognition from a large linguistically unconstrained time-spaced data base. IEEE Trans Acoust Speech Signal Process 27(1):74–82
53. Beigi H (2009) Effects of time lapse on speaker recognition results. In: 16th International Conference on Digital Signal Processing, 2009. IEEE, pp 1–6
54. Beigi H (2011) Fundamentals of speaker recognition. Springer Science & Business Media
55. Lamel LF, Gauvain JL (2000) Speaker verification over the telephone. Speech Commun 31 (2):141–154
56. Kelly F, Harte N (2011) Effects of long-term ageing on speaker verification. In: European Workshop on Biometrics and Identity Management. Springer, Berlin, Heidelberg, pp 113–124
57. Kelly F, Drygajlo A, Harte N (2012) Speaker verification with long-term ageing data. In: 2012 5th IAPR International Conference on Biometrics (ICB). IEEE, pp 478–483
58. Wang L, Wang J, Li L et al (2016) Improving speaker verification performance against long-term speaker variability. Speech Commun 79:14–29
59. Wang L, Zheng TF (2010) Creation of time-varying voiceprint database. In: Proc. Oriental-COCOSDA

Chapter 4
Application-Oriented Robustness Issues

4.1 Application Scenarios

With the development of speaker recognition technologies, they have been used in wide application areas. The main applications of speaker recognition technologies include the followings.

4.1.1 User Authentication

User authentication is one of the most popular applications of biometric recognition. Speaker recognition could be used in commercial transactions as an authentication method combined with some other techniques like password or face recognition. More recent applications are in user authentication for remote electronic and mobile payment [1]. Alternatively, it could be used for access control for computer login, such as a "key" to a physical facility, or in border control [2].

4.1.2 Public Security and Judicature

Another application field is the public security and judicature for law enforcement, including parolees monitoring (call parolees at random times to verify if they are in the restricted area) and prison call monitoring (validate inmates prior to outbound call) [3]. Speaker recognition technologies have also been used for forensics analysis (proving the identity of a recorder voice to convict a criminal or discharge an innocent in court) [4].

© The Author(s) 2017
T.F. Zheng and L. Li, *Robustness-Related Issues in Speaker Recognition*,
SpringerBriefs in Signal Processing, DOI 10.1007/978-981-10-3238-7_4

4.1.3 Speaker Adaptation in Speech Recognition

The speaker variability is one of the major problems in speech recognition, however on the contrary it can be an advantage in speaker recognition. Thus, speaker recognition technologies could be used to reduce the speaker variability in speech recognition systems by speaker adaptation (the system adapts its speech recognizer parameters to suit better for the current speaker, or to select a speaker-dependent speech recognizer from its database) [5].

4.1.4 Multi-speaker Environments

In a multi-speaker environment, transcribing *"who spoke when"* is one of the most basic requirements for dealing with audio recordings, such as recorded meetings or the audio portion of broadcast shows [6]. There are three different kinds of multi-speaker tasks: speaker detection, speaker tracking, and speaker segmentation [7].

4.1.5 Personalization

The voice-web, voice-mail or device customization is becoming more and more popular due to the development of speech technology. The speaker recognition technique can be used to identify whether the incoming voice is from a known speaker so that the system can be adapted according to this specific speaker's needs and preferences.

The above applications require robust speaker recognition technologies. Although, in recent years, the performance of speaker recognition has been considerably high under some restricted conditions, there are still many practical issues to face, including noisy variability, channel variability and intra-individual variations as mentioned above. All these variations lead to the mismatch between training and test that finally causes severe performance degradations. Therefore, robustness is one of the critical factors that impacts the success of speaker recognition in these applications [8].

4.2 Short Utterance

For speaker recognition using statistical methods, a sufficiently large amount of speech data for both model training and utterance recognition is necessary. This becomes a bottleneck for practical applications because in real scenarios it is often difficult for the system to collect long utterances (for example in forensic applications) or the user mostly does not prefer to speak too long (for example in e-banking). In this situation, the short utterance speaker recognition (SUSR) becomes a hot research topic.

If the enrollment and test utterances contain the same phone sequence (so called 'text-dependent' task), short utterances would not be a big problem [9]; however for text-independent tasks, severe performance degradation is often observed if the enrollment/test utterances are not long enough, as has been reported in several previous studies. For instance, Vogt et al. [10] reported that when the test speech was shortened from 20s to 2s, the performance degraded sharply in terms of Equal Error Rate (EER) from 6.34 to 23.89% on a NIST SRE task. Mak et al. [11] showed that when the length of the test speech is less than 2s, the EER raised to as high as 35.00%.

Research in [12] argued that the difficulty associated with text-independent SUSR can be largely attributed to the mismatched distributions of speech data between enrollment and test. Assuming that the enrollment speech is sufficiently long, the speaker model can be well trained. If the test speech is sufficient as well, the distribution of the test data tends to match the distribution represented by the speaker model; however, if the test speech is short, then only a part of the probability mass represented by the speaker model can be covered by the test speech. For a GMM-UBM system, this is equal to say that only a few Gaussian components of the model are covered by the test data, and therefore the likelihood evaluation is biased. For the i-vector model, since the Gaussian components share some statistics via a single latent variable, the impact of short test speech is partly alleviated. However, the limited data anyway leads to insufficient evaluation of the Baum-Welch statistics, resulting in a less reliable i-vector inference.

Research on short utterance speaker recognition (SUSR) is still very poor. In [13], the authors showed that performance on short utterances could be improved through the JFA framework that models the speaker and channel variabilities in two separate subspaces. This work was extended in [14] which reported that the i-vector model could distill speaker information in a more effective way so it was more suitable for SUSR. In addition, a score-based segment selection technique was proposed in [15], which evaluated the reliability of each test speech segment based on a set of cohort models, and scored the test utterance with the reliable segments only. A relative EER reduction of 22.00% was reported by the authors on a recognition task where the test utterances were shorter than 15s in length.

In [16], Li et al. studied two phonetic-aware approaches to text-independent SUSR tasks. The main idea was to resort the speech recognition to identify the phone content of speech signal, and then build speaker models on individual phones. These approaches could be regarded as a transfer from a text-independent SUSR task to a text-dependent one. Experimental results showed that both of these two phonetic-aware approaches could be superior to their respective baselines.

4.3 Anti-spoofing

It is widely aware of that most biometric systems are vulnerable to spoofing, also known as imposture. As early as in the 1990s, the anti-spoofing issue of biometric technologies had been focused on by researchers. For the fingerprint and face recognition, a series of methods and countermeasures have been studied to prevent different kinds of spoofing attacks. With the rapid development and wide application of speaker recognition technology, the study of speaker recognition on anti-spoofing issue also raised. The scenario of spoofing attacks on speaker recognition can be divided into four classes [17]: impersonation, speech synthesis, voice conversion, and replay. There are three special sessions on spoofing and countermeasures for automatic speaker verification held during INTERSPEECH 2013 [17], 2015 [18] and 2017 [19]. The first edition in 2013 was targeted mainly at increasing awareness of the spoofing problem. The 2015 edition included a first challenge on the topic, with commonly defined evaluation data, metrics and protocols. The task in ASVspoof 2015 was to discriminate genuine human speech from speech produced using text-to-speech (TTS) and voice conversion (VC) attacks. The primary technical goal of ASVspoof 2017 will be to assess spoofing attack detection accuracy with 'out in the wild' conditions, thereby advancing research towards generalized spoofing countermeasure, in particular to detect replay.

This section reviews past work to evaluate vulnerabilities and to develop spoofing countermeasure.

4.3.1 Impersonation

Impersonation refers to spoofing attacks with human-altered voices and is one of the most obvious forms of spoofing and earliest studied [17].

The work in [20] showed that non-professional impersonators can readily adapt their voice to overcome ASV, but only when their natural voice is already similar to that of the target. Further work in [21] showed that impersonation increased FAR rates from close to 0% to between 10 and 60%, but no significant difference in vulnerability to non-professional or professional impersonators.

None of the above studies investigated countermeasures against impersonation. Impersonation involves mostly the mimicking of prosodic or stylistic cues rather than those aspects more related to the vocal tract. Impersonation is therefore considered more effective in fooling human listeners than a genuine threat to today's state-of-the-art ASV systems [22]. With the development of speaker recognition technology, the impersonation attack is believed to be the easiest to detect.

4.3.2 Speech Synthesis

In recent years, the technology of speech synthesis has been developing rapidly. It can be made a specific-speaker adaptation with only a small amount of speech, and then accomplish spoofing attack to speaker recognition systems. There is a considerable volume of research in the literature which has demonstrated the vulnerability of ASV to synthetic voices.

ASV vulnerabilities to synthetic speech were first demonstrated over a decade ago [23] using an HMM-based, text-prompted ASV system [24] and an HMM-based synthesizer where acoustic models were adapted to specific human speakers [25, 26]. The ASV system scored feature vectors against speaker and background models composed of concatenated phoneme models. When tested with human speech the ASV system achieved an FAR of 0% and a false rejection rate (FRR) of 7%. When subjected to spoofing attacks with synthetic speech, the FAR increased to over 70%, however this work involved only 20 speakers.

Previous work has demonstrated the successful detection of synthetic speech based on prior knowledge of the acoustic differences of specific speech synthesizers, such as the dynamic ranges of spectral parameters at the utterance level [27] and variance of higher order parts of mel-cepstral coefficients [28]. There are some attempts which focus on acoustic differences between vocoders and natural speech. Since the human auditory system is known to be relatively insensitive to phase, vocoders are typically based on a minimum-phase vocal tract model. This simplification leads to differences in the phase spectra between human and synthetic speech, differences which can be utilized for discrimination [29, 30]. Other approaches to synthetic speech detection use F0 statistics [31, 32], based on the difficulty in reliable prosody modeling in both unit selection and statistical parametric speech synthesis. F0 patterns generated for the statistical parametric speech synthesis approach tend to be over-smoothed and the unit selection approach frequently exhibits 'F0 jumps' at concatenation points of speech units.

4.3.3 Voice Conversion

Voice conversion is a sub-domain of voice transformation [33] which aims to convert one speaker's voice towards that of another [33]. The field has attracted increasing interest in the context of ASV vulnerabilities for over a decade [34].

The process of voice conversion generally makes up of two steps: off-line training and on-line conversion. Therefore, how to establish the conversion function between anti-spoofing speech and target speech determines the effect of spoofing attack.

Nowadays, there are some methods to detect converted voice. While the proposed cos-phase and modified group delay function (MGDF) phase countermeasures proposed in [30] are effective in detecting synthetic speech, they are unlikely to detect converted voice with real-speech phase [35]. Two approaches to artificial signal detection were reported in [36]. Experiments showed that supervector-based SVM classifiers could be naturally robust to such attacks whereas all spoofing attacks could be detected using an utterance-level variability feature which detects the absence of natural, dynamic variability characteristic of genuine speech. An alternative approach based on voice quality analysis was regarded to be less dependent on explicit knowledge of the attack but less effective in detecting attacks. A related approach to detect converted voice was proposed in [37]. Probabilistic mappings between source and target speaker models were shown to yield converted speech with less short-term variability than genuine speech. The threshold which was the average pair-wise distance between consecutive feature vectors was used to detect converted voice with an EER lower than 3%.

4.3.4 Replay

Replay attacks [38] involve the presentation of speech samples captured from a genuine client in the form of continuous speech recordings, or samples resulting from the concatenation of shorter segments.

A replay attack is illustrated in Fig. 4.1 [39]. An attacker first acquires a covert recording of the target speaker's utterance without his/her consent, and then replays it using a replay device in some physical environment. The replayed speech is captured by the ASV system terminal (recording device). In contrast, an authentic target speaker recording illustrated in the lower part of Fig. 4.1 would also be obtained through some (generally, another) physical space, but captured directly by the ASV system mic.

Compared with the former three spoofing scenarios, the implementation of replay attacks requires no specific expertise nor any sophisticated equipment and thus they arguably present a greater risk. Replay attack is much easier to occur in practical applications. The intruder was able to break into the voiceprint recognition system without the need of speech processing techniques, and only required a high-quality recording and playback devices.

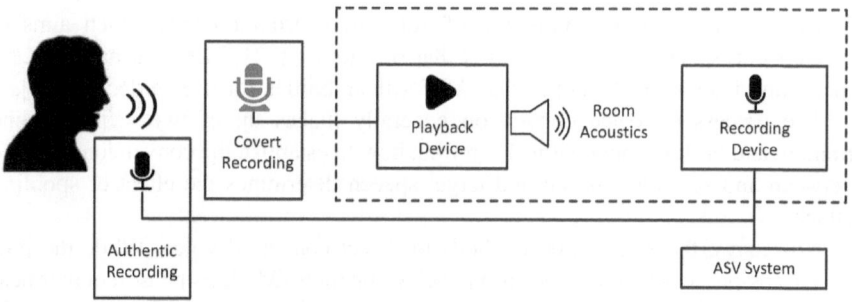

Fig. 4.1 An illustration of replay spoofing

Researches have been demonstrated that low-effort replay attacks provoke higher levels of false acceptance than comparatively higher-effort spoofing attacks such as voice conversion and speech synthesis.

Work in [40] investigated vulnerabilities to the replaying of far-field recorded speech. Using a baseline ASV system based on JFA, their work showed an increase in the equal error rate (EER) from 1% to almost 70% when imposter accesses were replaced by replayed spoof attacks. The same authors showed that it was possible to detect such spoofing attacks by measuring the channel differences caused by far-field recording [41]. While they show spoof detection error rates of less than 10% it was feasible that today's state-of-the-art approaches to channel compensation would leave some systems more vulnerable to replay attacks.

Reference [42] re-assessed the threat of replay attacks against automatic speaker verification (ASV) systems. The work was performed using simulated attacks and with large, standard NIST speaker recognition corpora and six ASV systems. Experimental results showed that low-effort replay attacks pose a significant risk, and the state-of-the-art iVector-PLDA system was the most vulnerable to replay attacks, especially for FARs below 10%.

Considering the threat of the replay attack, researchers carried out a series of studies. Reference [43] evaluated the vulnerability of text-dependent speaker verification systems under the replay attack using a standard benchmarking database, and also propose an anti-spoofing technique to safeguard the speaker verification systems. The key idea of the spoofing detection technique is to decide whether the presented sample is matched to any previous stored speech samples based a similarity score. In [44], the authors focused on detecting if the test segment was a far-filed microphone recording of the victim. They assumed that a far-field recording will cause an increment of the noise and reverberation levels of the signal. This would have as consequence a flattening of the spectrum and a reduction of the modulation indexes of the signal. Therefore, several features were selected to detect the far-field replay attack.

4.3.5   Voice Liveness Detection

In addition to these mentioned solutions, voice liveness detection is also one kind of effective anti-spoofing mechanism. Some researchers [45] expected that the acoustic differences between artificial and natural speech to gradually become smaller and eventually marginal in the near future. Looking at other biometrics fields, such as face, fingerprint and even iris recognition systems, the most effective countermeasures for spoofing attacks is to use a 'liveness detection' framework that ensures that the person attempting authentication is alive. The task of voice liveness detection is to validate whether the presented speech signals was originated from a live human. In [45, 46], a pop noise detection algorithm was proposed and experimental results showed that this algorithm can be used to discriminate live voice signals from artificial ones generated by means of speech synthesis techniques.

4.4   Cross Encoding Schemes

In transmission, speech signals are often encoded so as to increase the efficiency. However, it will distort the signal to some extent. Depending on the task at the receiving end, the distortion could be acceptable or not. For example, the bandwidth of the telephone channel is mostly 3.4 kHz, and either the A-law or the μ-law encoding scheme can reserve enough speech content information which is acceptable for a speech recognition task to some extent, yet relatively insufficient for a speaker recognition task because more speaker information resides in high frequency end. In spite of this, if there is no training and testing mismatch, the problem is not so severe especially if there is enough data involved where there are rich kinds of information to use.

The situation is in recent application, several different kinds of encoding schemes are being popular used, such as A-law/μ-law ADPCM, G.711, G.728, and WeChat specific format. The encoding scheme mismatch between test and training can lead to a severe performance degradation. Few research is focused on this topic which must be paid more attention to. But on the contrary, with the development of telecommunication technologies (such as 4G/5G), if the encoding rate is not so critical and is high enough for speech signals, the problem may become less important.

4.5   Discussion

In this chapter, we first summarize five application scenarios of speaker recognition technologies, and then overview three application-oriented robustness issues, they are short utterance, anti-spoofing and cross encoding. Along with the speaker recognition of extensive application, these application-oriented issues will become the research hotspot and more potential issues will be appeared.

References

1. Reynolds DA (1995) Automatic speaker recognition using Gaussian mixture speaker models In: The Lincoln Laboratory Journal
2. Kinnunen T (2003) Spectral features for automatic text-independent speaker recognition. Licentiate's Thesis, University of Joensuu—2003
3. Reynolds D, Heck LP (2001) Speaker verification: from research to reality. In: Tutorial of International Conference on Acoustics, Speech, and Signal Processing

4. Rose P (2003) Forensic speaker identification. CRC Press
5. Kuhn R, Junqua JC, Nguyen P et al (2000) Rapid speaker adaptation in eigenvoice space. IEEE Trans Speech Audio Process 8(6):695–707
6. Dunn RB, Reynolds DA, Quatieri TF (2000) Approaches to speaker detection and tracking in conversational speech. Digit Signal Proc 10(1–3):93–112
7. Martin AF, Przybocki MA (2001) Speaker recognition in a multi-speaker environment. In: INTERSPEECH. pp 787–790
8. Jin Q (2007) Robust speaker recognition. Carnegie Mellon University
9. Larcher A, Lee KA, Ma B et al (2012) RSR2015: database for text-dependent speaker verification using multiple pass-phrases. In: INTERSPEECH. pp 1580–1583
10. Vogt R, Sridharan S, Mason M (2010) Making confident speaker verification decisions with minimal speech. IEEE Trans Audio Speech Lang Process 18(6):1182–1192
11. Mak MW, Hsiao R, Mak B (2006) A comparison of various adaptation methods for speaker verification with limited enrollment data. In: Proceedings of the 2006 IEEE International Conference on Acoustics, Speech and Signal Processing, 2006, ICASSP 2006, vol 1. IEEE, p I-I
12. Li L, Wang D, Zhang C et al (2016) Improving short utterance speaker recognition by modeling speech unit classes. IEEE/ACM Trans Audio Speech Lang Process (TASLP) 24 (6):1129-1139
13. Vogt RJ, Lustri CJ, Sridharan S (2008) Factor analysis modelling for speaker verification with short utterances
14. Kanagasundaram A, Vogt R, Dean DB et al (2011) I-vector based speaker recognition on short utterances. In: Proceedings of the 12th Annual Conference of the International Speech Communication Association. International Speech Communication Association (ISCA), pp 2341–2344
15. Nosratighods M, Ambikairajah E, Epps J et al (2010) A segment selection technique for speaker verification. Speech Commun 52(9):753–761
16. Li L, Wang D, Zhang X et al (2016) System combination for short utterance speaker recognition. arXiv preprint arXiv:1603.09460
17. Evans NWD, Kinnunen T, Yamagishi J (2013) Spoofing and countermeasures for automatic speaker verification. In: Interspeech. pp 925–929
18. Wu Z, Kinnunen T, Evans N et al (2015) ASVspoof 2015: the first automatic speaker verification spoofing and countermeasures challenge. Training 10(15):3750
19. Kinnunen T, Evans N, Yamagishi J et al (2017) ASVspoof 2017: automatic speaker verification spoofing and countermeasures challenge evaluation plan. http://www.spoofingchallenge.org/
20. Lau YW, Wagner M, Tran D (2004) Vulnerability of speaker verification to voice mimicking. In: Proceedings of the 2004 International Symposium on Intelligent Multimedia, Video and Speech Processing, 2004. IEEE, pp 145–148
21. Lau YW, Tran D, Wagner M (2005) Testing voice mimicry with the YOHO speaker verification corpus. In: International Conference on Knowledge-Based and Intelligent Information and Engineering Systems. Springer, Berlin, Heidelberg, pp 15–21
22. Lindberg J, Blomberg M (1999) Vulnerability in speaker verification-a study of technical impostor techniques. In: Eurospeech 99:1211–1214
23. Masuko T, Hitotsumatsu T, Tokuda K et al (1999) On the security of HMM-based speaker verification systems against imposture using synthetic speech. In: Eurospeech
24. Matsui T, Furui S (1995) Likelihood normalization for speaker verification using a phoneme-and speaker-independent model. Speech Commun 17(1):109–116
25. Masuko T, Tokuda K, Kobayashi T et al (1996) Speech synthesis using HMMs with dynamic features. In: 1996 IEEE International Conference on Acoustics, Speech, and Signal Processing, ICASSP 1996, vol 1. IEEE, pp 389–392
26. Masuko T, Tokuda K, Kobayashi T et al (1997) Voice characteristics conversion for HMM-based speech synthesis system. In: 1997 IEEE International Conference on Acoustics, Speech, and Signal Processing, 1997, ICASSP 1997, vol 3. IEEE, pp 1611–1614

27. Satoh T, Masuko T, Kobayashi T et al (2001) A robust speaker verification system against imposture using an HMM-based speech synthesis system. In: INTERSPEECH. pp 759–762
28. Chen LW, Guo W, Dai LR (2010) Speaker verification against synthetic speech. In: 2010 7th International Symposium on Chinese Spoken Language Processing (ISCSLP). IEEE, pp 309–312
29. De Leon PL, Pucher M, Yamagishi J et al (2012) Evaluation of speaker verification security and detection of HMM-based synthetic speech. IEEE Trans Audio Speech Lang Process 20 (8):2280–2290
30. Wu Z, Siong CE, Li H (2012) Detecting converted speech and natural speech for anti-spoofing attack in speaker recognition. In: INTERSPEECH. pp 1700–1703
31. Ogihara A, Hitoshi U, Shiozaki A (2005) Discrimination method of synthetic speech using pitch frequency against synthetic speech falsification. IEICE Trans Fundam Electron Commun Comput Sci 88(1):280–286
32. De Leon PL, Stewart B, Yamagishi J (2012) Synthetic speech discrimination using pitch pattern statistics derived from Image analysis. In: INTERSPEECH. pp 370–373
33. Stylianou Y (2009) Voice transformation: a survey. In: IEEE International Conference on Acoustics, Speech and Signal Processing, 2009, ICASSP 2009. IEEE, pp 3585–3588
34. Pellom BL, Hansen JHL (1999) An experimental study of speaker verification sensitivity to computer voice-altered imposters. In: Proceedings of the 1999 IEEE International Conference on Acoustics, Speech, and Signal Processing, 1999, vol 2. IEEE, pp 837–840
35. Matrouf D, Bonastre JF, Fredouille C (2006) Effect of speech transformation on impostor acceptance. In: Proceedings of the 2006 IEEE International Conference on Acoustics, Speech and Signal Processing, 2006, ICASSP 2006, vol 1. IEEE, p I-I
36. Alegre F, Vipperla R, Evans N (2012) Spoofing countermeasures for the protection of automatic speaker recognition systems against attacks with artificial signals. In: 13th Annual Conference of the International Speech Communication Association, INTERSPEECH 2012
37. Alegre F, Amehraye A, Evans N (2013) Spoofing countermeasures to protect automatic speaker verification from voice conversion. In: 2013 IEEE International Conference on Acoustics, Speech and Signal Processing (ICASSP). IEEE, pp 3068–3072
38. Lindberg J, Blomberg M (1999) Vulnerability in speaker verification-a study of technical impostor techniques. In: Eurospeech, vol 99. pp 1211–1214
39. Kinnunen T, Sahidullah M, Falcone M et al (2017) RedDots replayed: a new replay spoofing attack corpus for text-dependent speaker verification research. In: Proceedings of the IEEE International Conference on Acoustics, Speech and Signal Processing
40. Villalba J, Lleida E (2010) Speaker verification performance degradation against spoofing and tampering attacks. In: FALA workshop. pp 131–134
41. Villalba J, Lleida E (2011) Preventing replay attacks on speaker verification systems. In: 2011 IEEE International Carnahan Conference on Security Technology (ICCST). IEEE, pp 1–8
42. Alegre F, Janicki A, Evans N (2014) Re-assessing the threat of replay spoofing attacks against automatic speaker verification. In: 2014 International Conference of the Biometrics Special Interest Group (BIOSIG). IEEE, pp 1–6
43. Wu Z, Gao S, Cling ES et al (2014) A study on replay attack and anti-spoofing for text-dependent speaker verification. In: 2014 Annual Summit and Conference on Asia-Pacific Signal and Information Processing Association (APSIPA). IEEE, pp 1–5
44. Villalba J, Lleida E (2011) Detecting replay attacks from far-field recordings on speaker verification systems. In: European Workshop on Biometrics and Identity Management. Springer, Berlin, Heidelberg, pp 274–285
45. Shiota S, Villavicencio F, Yamagishi J et al (2015) Voice liveness detection algorithms based on pop noise caused by human breath for automatic speaker verification. In: INTERSPEECH. pp 239–243
46. Shiota S, Villavicencio F, Yamagishi J et al (2016) Voice liveness detection for speaker verification based on a tandem single/double-channel pop noise detector. Odyssey 2016:259–263

Chapter 5
Conclusions and Future Work

This book presents an overview of speaker recognition technology with an emphasis on how to deal with the robustness issues. Firstly, we give an introduction of speaker recognition, including its basic concept, system framework, development history, categories and performance evaluations. And then we summarize the speaker recognition robustness issues from three categories, which are environment-related, speaker-related, and application-oriented. For each category, we review the current hot topics, existing technologies, and foresee potential research focuses in the future.

Currently, although speaker recognition technology is suffering from many technical difficulty problems and still has a certain distance from the practical applications, it is obvious that there exists a enormous potential from numerous aspects in the future. It is expected to be applied to access control, transaction authentication, voice-based information retrieval, forensic analysis, personalization of user devices, etc. Besides, in practical applications, the speaker recognition technology can also be combined with other biometric recognition technologies, such as face, fingerprint, and irise, to enhance the system security. In the remote identity authentication, speaker recognition systems supplemented with manual work can accomplish the double-authentication. In the mobile payment and voiceprint lock verification, based on the dynamic random verification code and automatic speech recognition, speaker recognition systems are able to prevent some spoofing attacks. In conclusion, according to various practical requirements, speaker recognition technologies can be applied in a flexible way.

© The Author(s) 2017
T.F. Zheng and L. Li, *Robustness-Related Issues in Speaker Recognition*,
SpringerBriefs in Signal Processing, DOI 10.1007/978-981-10-3238-7_5